T0358384

Graph Coloring

This book explores the problem of minimal valid graph coloring, first in the form of games and then of resolution algorithms. Emphasis is placed on deterministic, guaranteed, and non-guaranteed methods. Stochastic methods are then just mentioned because they are already widely described in previous publications.

The study then details a general quantum algorithm of polynomial complexity. A final chapter provides elements of reflection on diplomatic algorithms that, for the problem of coloring under resource constraints, seek a compromise minimizing frustrations. The appendix includes some mathematical additions and the source codes of the main algorithms presented, in particular the one of the quantum method.

Maurice Clerc is a French mathematician. He worked in the Research and Development Department of France Télécom. Retired since 2004, he remains active in various fields of research, both in particle swarm optimization (PSO)-related fields and in other areas. He regularly publishes articles, gives keynote speeches in conferences, serves as a director and thesis jury, and occasionally works as an optimization consultant.

Advances in Metaheuristics

Series Editors:
Patrick Siarry, *Universite Paris-Est Creteil, France*
Anand J. Kulkarni, *Symbiosis Center for Research and Innovation, Pune, India*

This series delves into the topic of AI-based Metaheuristics and explores the latest advances in this area.

For more information about this series please visit:
https://www.routledge.com/Advances-in-Metaheuristics/book-series/AIM

Graph Coloring

From Games to Deterministic and Quantum Approaches

Maurice Clerc

CRC Press
Taylor & Francis Group
Boca Raton London New York

CRC Press is an imprint of the
Taylor & Francis Group, an **informa** business

First edition published 2025
by CRC Press
2385 NW Executive Center Drive, Suite 320, Boca Raton FL 33431

and by CRC Press
4 Park Square, Milton Park, Abingdon, Oxon, OX14 4RN

CRC Press is an imprint of Taylor Francis Group, LLC © 2025 Maurice Clerc

ISBN: 978-1-032-73751-5 (hbk)
ISBN: 978-1-032-76266-1 (pbk)
ISBN: 978-1-003-47778-5 (ebk)

DOI: 10.1201/ 9781003477785

Typeset in Minion
by Deanta Global Publishing Services, Chennai, India

TLe jeu est l'occupation la plus sérieuse de l'enfant
(Playing is the most serious occupation of the child)

Albert Brie (Le mot du silencieux)

Contents

Preface

In nursery and primary schools, children are sometimes amused by having to color a drawing made of circles and lines between the circles, with two rules:

- two circles connected by a line must not be of the same color;

- the best coloring is the one that uses the fewest different colors.

It's interesting to see how the children try to find the best solution, especially when they can use colorful tokens and move them around instead of using colored pencils.

However, the issue, which falls under the deep topic of coloring graphs in the best way, can be quite complicated. This little book doesn't claim to cover everything about it. Instead, I want to show different levels of difficulties, starting from what might seem like easy games to more complex things like quantum algorithms.

Between these two extremes, I particularly emphasize deterministic resolution methods as they appear to be undervalued compared to those using randomness. As for the latter, of which I provide only a brief overview, there exists an extensive body of literature that interested readers can readily explore through searches using keywords such as *graph coloring* and *stochastic methods*.

You can learn more about this topic from books, articles, or online searches. These resources cover different ways of graph coloring and study specific types of graphs, like planar graphs (which can be drawn without lines crossing) and trees. For example, a planar graph needs only four colors at most, and a tree graph needs only two.

Since any presentation of a scientific work should provide the essential elements for potential replication, I have included the most crucial source codes that were used.

Finally, the objective of this book is to make you think about interesting graph coloring problems. Some coding definitions are given without examples on purpose, so that you can explore and understand them yourself. Similarly, some statements are made without exhaustive proof, although all the necessary elements for such proofs are readily available.

Wishing you an enjoyable reading experience and fruitful reflections!

TOOLS

A substantial portion of the documentation for this book was acquired through online research via the Startpage metasearch engine (https://startpage.com). Its primary advantage lies in its capability to simultaneously submit the same query to multiple search engines and aggregate the results, all while divulging the bare minimum of information about the query's origin to the search engines.

The writing process was facilitated by LyX (http://lyx.org), a tool that simplifies the generation of LaTeX files without requiring proficiency in the LaTeX language. Bibliography management was carried out using Zotero (https://www.zotero.org) in conjunction with its LyZ extension for seamless integration with LyX. The figures were created using the following software packages:

LibreOffice Calc (http://www.libreoffice.org);

Gimp (http://www.gimp.org);

Inkscape (https://inkscape.org/);

GNU Octave (https://www.gnu.org/software/octave), which is a near-identical counterpart to the commercial software MATLAB®, and also MATLAB® itself.

Mathematica® (https://www.wolfram.com/mathematica/).

The computer programs were authored in either MATLAB® or Octave. The computational processing was ultimately executed on a 64-bit microcomputer, operating under the Ubuntu 22.04 Linux platform. It's worth noting that the epsilon machine for this system is 2.22×10^{-16}, which means that any computation involving absolute values assumed to be less than this threshold should be approached with the utmost caution.

CONTACTING THE AUTHOR

If you wish to share comments, report errors, or offer suggestions, please feel free to reach out to me via email at Maurice.Clerc@WriteMe.com.

THANKS

To Patrick Siarry who consistently supported me on this project, even though it wasn't very orthodox.

To Abhi Dattasharma, who effectively corrected my translation from the original French version.

To these two friends: thank you!

And thanks to the children who contributed, sometimes without knowing it, to improving certain games, especially the Colorigraph and the Quantum Race. If they ever read this book, they will recognize themselves.

## CHAPTER 1

Games

This chapter might seem unimportant at first, but a close look at the three games described below, especially the last two, which are relatively more complicated, reveals that those are interesting to watch and it is worthwhile to understand the strategies used. In fact, most of the algorithms we'll talk about later are based on these kinds of strategies.

Let's take a simple look at the graph in Figure 1.1 and start explaining a bit. The circles are called *vertices*, and the lines between them are edges. People also use the words "nodes" and "arcs." However, the term "arc" is usually used when the lines have arrows, which can be seen as showing an influence or implication.

For each node we can count the number of edges attached to it: it is the degree of the node. In Figure 1.1, we see that there is one node of degree six, three of degree five, two of degree four, and one of degree three. The sum of degrees is 32, and the number of edges is therefore half, 16.

Here, our interest lies in the various ways of coloring the nodes of the graph while imposing rules and constraints. They can be considered as games, and some of them are challenging (Guignard 2011).

Note that this chapter is the first of the progression from "easy" (games) toward "difficult" (quantum algorithms) and can therefore be skipped if you are already familiar with the topic.

1.1 COL'S GAME

It is usually practiced on a planar graph, but this is not mandatory. Note that despite appearances, the graph in Figure 1.1 is planar, which can be shown by redrawing it differently (Figure 1.2, where, for easy understanding, the nodes have been numbered). It is not necessary for the edges to be straight, as they simply represent a connection between two nodes.

1.1.1 Rules

The game of Col can be practiced from six years of age, and by even younger children after some modifications of the words and presentation of the rules. The name is both an allusion to the creator of the game (Colin Vout) and the fact that it is a coloring game.

The most common form of it is a two-player game, with two colored pencils, for example, "blue" 🖊 and "red" ⬤ .

DOI: 10.1201/9781003477785-1

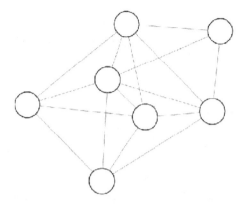

FIGURE 1.1   A seven nodes and sixteen edges graph.

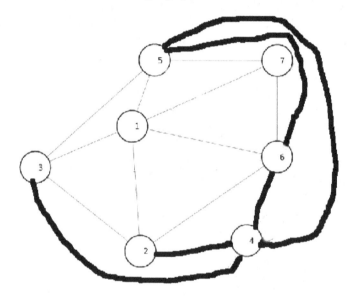

FIGURE 1.2   We can redraw the graph in Figure 1.1 without crossing edges: it is planar.

Let us call the player with the blue pencil B, and the player with the red pencil R. B begins the game. B chooses any node and colors it. Then, it is R's turn. R colors one of the remaining nodes. Then, B colors one of the remaining nodes, and so on. But you should never color a node in the same color as one of the neighboring nodes (to which it is adjacent, that is to say connected by a line).

It was perhaps clear from the discussion, but let us note specifically that we consider only connected graphs, that is to say for which we can always go from one node to another by following the edges, bidirectionally (an edge can be traversed in both directions) and without any loop (no edge from a node to itself).

The first one who is stuck (because all the nodes not yet colored are adjacent to a node of the color of the pencil held by the player whose turn it is) loses.

On large graphs, you can play with more than two people and with more colors.

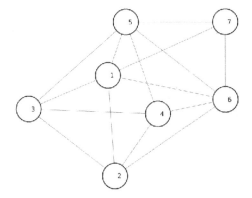

FIGURE 1.3 Between nodes 3 and 7, the distance is 2, following the path 3-1-7. In this graph there is also no distance greater than 2, which is the value of its diameter.

1.1.2 Strategies

Even for this simple game, there is no known method to win for sure,[1] but there are still a few strategies that often work. Let us first define the *distance* between two nodes A and B: this is the number of edges of the shortest path from A to B. When considering distances for all pairs of nodes, the largest is called the *diameter* of the graph. In Figure 1.3, it is two, even if it is not obvious. We then have as rules of thumb:

- color a node of the smallest degree possible;

- if there are several candidates, choose one that is as far away as possible from those already colored with the current color.

Note also the rule of (even, lose)[2]: if the diameter is even, then the player who starts risks losing, if the opponent plays well.

1.1.3 Outside

A variation of Col's game is presented in Dorian Mazauric's book on "life size" graph games (Mazauric 2016), even if it is not explicitly stated. A graph is drawn on the ground and teams compete, each of a given color (shits, scarf, armband, etc.). Each team places one of its members on a free (vacant) node of the graph.

The first team that is forced to place a player "next to" another player of the same color has lost. Here, "next to" means of course that there is an edge connecting the two positions. A player, once placed, cannot be moved. Teams must therefore apply a "greedy" strategy, as described below in the section 4.2 (Figure 1.5).

---

[1] Written in November 2023.
[2] Sounds better in French: *pair perd*

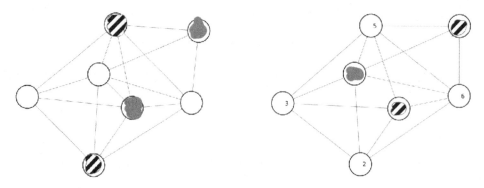

FIGURE 1.4 Game of Col. A game ending. In the first game, the "blue" player starts badly and loses. In the second, the blue player starts better on a low-degree node, and wins, with the "red" player having wrongly played on node 1 instead of 3.

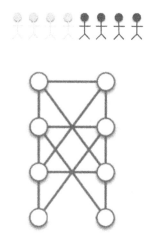

FIGURE 1.5 The game of Col per team outside. With this setup, if the team that starts plays well, it wins.

1.2 COLORIGRAPH

The Colorigraph is a game of strategy and speed! It can be played from the age of five or six, but make no mistake, even adults can sometimes find themselves stuck.

In this game, one needs to color graphs with 12 nodes, though the number, in principle, can be increased. However, even a 12-node game is often not simple, and because in this game we limit ourselves to three colors at most, there may not be a solution. But we have to be sure, so in case of doubt we can use a computer algorithm *a posteriori* to find out if a solution exists and to give a solution (perhaps even the best solution) if it exists.

Let us make a detailed description of this game, with examples.

The "deluxe" version requires a specific material (see the list below and Figure 1.10), but we can just draw the graphs on a sheet. Nevertheless, color tokens are still useful, to try different possibilities easily.

1.2.1 Material

- A board with 12 numbered vertical cylindrical studs arranged on a circle (or pre-printed cards).

- Something to materialize the edges. Stems with rings at the ends (six stems for each of the seven possible sizes). You can also use elastic bands, or just a pencil if you use sheets.

- 12 flat pierced bicolor tokens, such as blue on one side, red on the other.

- Six black pierced flat tokens.

- Two 12-sided dice (numbered) or two 6-sided (to be tossed twice). If four dice are available, two can be used to record the stake.

- Keep a record of the stakes and points earned (paper + pencil or bar with balls, for example).

- Hourglass (30 seconds, 1 minute, 2 minutes) or timer. A slightly stressful timer (ticking and final buzzer) increases the difficulty.

1.2.2 Conduct of a Game

Depending on the players, you can increase or decrease the time allotted for each problem. Players agree on the total number of points to accumulate to win, for example, at least 100 points for a two-player game or 70 points for a three-player game (see below for the calculation of points) or less if a faster game is desired. Each player, in turn (except in case of challenge, see below):

- builds a problem;

- attempts to resolve it within the time allowed.

1.2.3 Build a Problem

The player whose turn it is set stakes a number between 2 and 24. Note this stake. The player to the left throws the 12-sided dice as many times as this stake. If the same two numbers appear, throw again. Each time, materialize the corresponding edge; that is, place a stem (or stretch an elastic or draw a line) between the studs (the nodes) whose numbers are given by the dice. Note: if the roll of the dice restores an existing edge, redo it.

This phase is omitted if the graphs are pre-drawn on sheets (see variant 3 below).

1.2.4 Solve a Problem (Coloring)

The coloring consists in putting a token on each stud, within the allowed time. A coloring is valid if no edge has tokens of the same color at both ends. Set the timer to one or two minutes, depending on the level of the players. Note: Single studs (which do not support any edge) can be ignored.

Three possible cases:

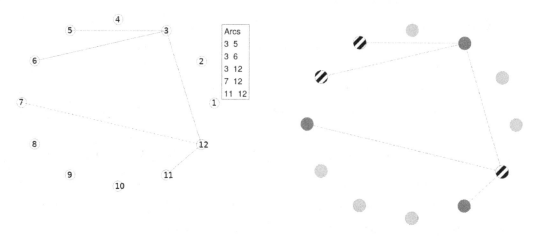

FIGURE 1.6   Sara's 5-stake. The problem and the solution found.

1. Valid coloring, only red and blue. The player wins a number of points equal to the stake.

2. Valid coloring but with black tokens. Each black token loses two points. The player thus wins the stake reduced by twice the number of black token used, unless there is a challenge (see below).

3. Invalid coloring. In cases 2 and 3 the other player (or one of the other two in the case of a three-way game) can throw a challenge, saying "I take!," "Challenge!," or "I can do better!"

1.2.5 Challenge

The player who has challenged (let us call this player as C) tries to solve the problem. As C had longer time to observe the graph, the timer is set to half the initial time. If, within the given time frame, C manages to find a better coloring, C gains points as indicated above.

For case 2, a better coloring is also a valid coloring but with fewer black tokens, or none.

For case 3, a better coloring is any valid coloring, with two or three colors.

Warning: if the player who launched the challenge does not succeed, the opponent wins the stake!

Note: after a challenge we continue the normal turn of the players. For a game with two people, the player who launched the challenge thus plays twice in a row.

1.2.6 Example of a Game Between Sara and Samuel

1. Sara is the youngest; she begins. She chooses a very careful five-stake. Samuel rolls the dice five times (5-3, 7-12, 11-12, 12-3, 6-3), and Sara puts the corresponding edge each time. The timer is set to one minute. She easily finds a solution and gains 5 points (see Figure 1.6).

2. Samuel's turn. He picks a slightly riskier 10-stake. Sara rolls the dice 10 times, and Samuel builds a 10-edge problem. The timer is set to one minute. Samuel does not find a two-color solution (which is impossible; see Figure 1.7), but with a black token, he still gets a valid coloring. He then wins 10 − 2 = 8 points.

3. To attempt to regain points, Sara tries a 12-stake. With 12 rolls of the dice (by Samuel), she builds a rather difficult problem. The timer is set to one minute. She does not find a two-color solution (it is impossible; see Figure 1.8). She then tries with black tokens, but the minute is past.

4. Samuel challenges. The timer is set to 30 seconds. Samuel finds a solution in time, with two black tokens. He wins $12 - 2 \times 2 = 8$ points, and it is up to him again. If he had not found a solution in time, Sara would have won 12 points.

5. Samuel tries a big shot: a 24-stake. He places the 24 edges defined by the dice thrown by Sara (this can be a bit long!). The timer is set to one minute. Samuel quickly sees that the bi-coloring is impossible (see Figure 1.9) and tries a tri-coloring using black tokens. But the minute has passed.

6. Sara, having observed the problem, poses a challenge. With the timer set to 30 seconds, she finds a four black tokens solution. She wins $24 + 4 \times 2 = 16$ points.

7. etc., until one of the players reaches at least the total number of points needed to win.

1.2.7 Variants

You can easily invent variants. Here are a few.

1.2.7.1 Variant 1

The stake is chosen at random by rolling the dice. Its value is given by the sum of the points.

1.2.7.2 Variant 2 (Simplification)

Build a problem (roll the dice, place the edges). Place 12 bi-color tokens on the 12 studs but with the same color visible. Start the timer. The resolution of the problem involves flipping certain tokens to achieve a valid bi-coloring. In the case of failure, we do not look for tri-coloring.

1.2.7.3 Variant 3 (with Additional Material)

Problems are predefined on cards. The deck of cards is mixed, placed on the table face down, and the player whose turn it is draws a card.

1.2.8 Some Tips

- If there are three edges forming a triangle, bi-coloring is not possible.

- If you see triangles, place a black token on the stud with the most edges.

- More generally, if you don't find a bi-coloring quickly, placing a black token on the stud with the most edges (maximum degree) is often a good strategy to find a valid coloring (but not necessarily the best one).

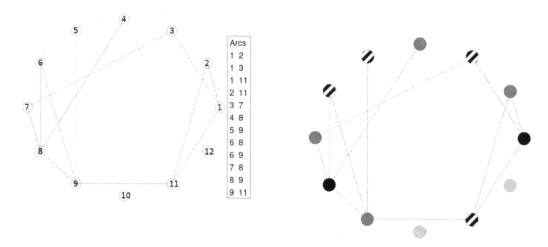

FIGURE 1.7 Samuel's 10-stake. The problem and the solution found, with a black token.

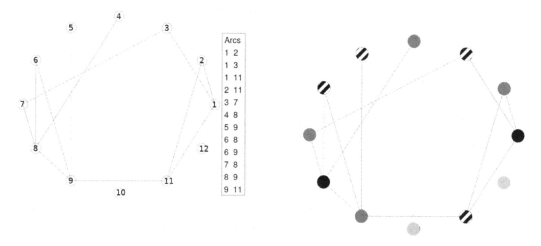

FIGURE 1.8 Sara's 12-stake. She does not find a solution in less than a minute, but Samuel challenges and finds one with two black tokens, in less than 30 seconds.

1.3 ALONE

Even though Colorigraph is usually played with two or three persons, the basic principle can be generalized for a single player: given a graph, color it with a minimum number of colors, the *chromatic number* of the graph. To generate and draw a graph, we can proceed in an analogous way, with dice:

- A first throw gives the number of nodes;

- A second throw gives the number of edges;

- The following throws give the origin and end numbers for each edge.

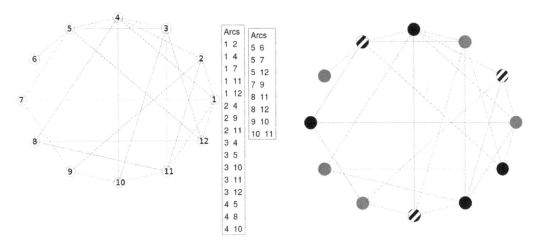

Arcs	Arcs
1 2	5 6
1 4	5 7
1 7	5 12
1 11	7 9
1 12	8 11
2 4	8 12
2 9	9 10
2 11	10 11
3 4	
3 5	
3 10	
3 11	
3 12	
4 5	
4 8	
4 10	

FIGURE 1.9   Samuel's 24-stake. But it is Sara who finds a solution with four black tokens, after having challenged. Note that there is a solution with only three black tokens, in 2, 5, and 10.

But it is much simpler to use a computer program for random generation of a connected bidirectional graph that you can display or print. A source code is given in the appendix 8.7.1.

Of course, we consider only graphs without loop, that is to say without any edge going from a node to itself, because otherwise the problem is obviously impossible.

Moreover, I invite you to actually try to solve such random graphs, by explaining the strategies implemented. Indeed, in the chapter on methods of resolution, you will see that most of these are just formalizations of intuitive strategies.

Figure 1.11 shows some examples of eight-node graphs and their most economical coloring in number of colors.

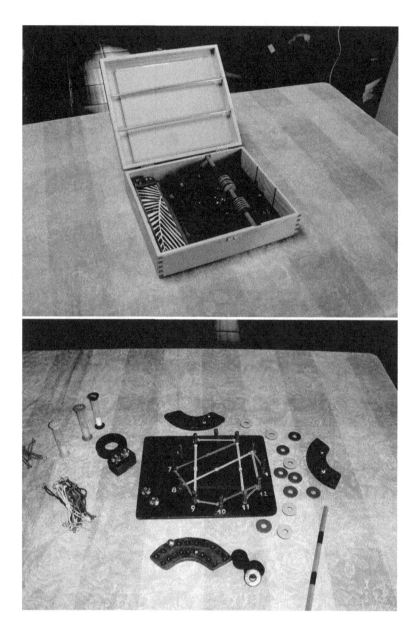

FIGURE 1.10   Colorigraph. Open box and a game in progress.

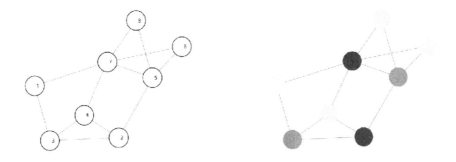

(a) Twelve edges. Planar graph, which needs only three colors.

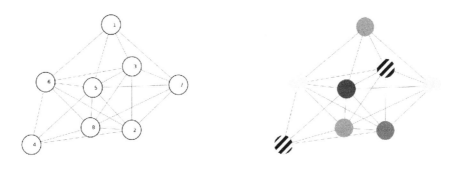

(b) With this 25 edges graph you need five colors

FIGURE 1.11   Examples of eight-node random connected graphs.

## CHAPTER 2

A Few Applications

Any allocation of resources with constraints of mutual exclusions (course schedule, frequency band sharing, scheduling of tasks to use the same machine, etc.) can be modeled by a graph and resolved by a valid coloring; that is to say, for each edge, the ends have different colors. There is always such a coloring, but the goal is also to use as few resources (colors) as possible.

Let's see some simple examples, essentially playful, just for illustration. The interested reader will easily find many more substantial ones in the literature (Ahmed 2022; Thadani et al. 2022; Lewis 2016).

2.1 CARPOOL

Alice, Bob, Samuel, Sara, and Nicolas must go to the same meeting. A carpool would be desirable, but Bob hates everyone; Samuel does not support Alice or Nicolas, who himself does not get along with Sara. What is the least number of vehicles needed?

To solve this problem, let us draw a graph whose nodes are our five characters and let us connect edges between incompatible individuals. We see that a valid coloring requires three colors. Red (rage?) for Bob, green for Alice and Nicolas, and blue for Sara and Samuel. So three vehicles are necessary and sufficient. Notice that Bob the misanthrope has to take a car for himself. What if there are only two cars available? A compromise will have to be found, and this more constrained problem is mentioned in Chapter 7, on diplomat algorithms (Figure 1.1).

Of course, here, a solution can easily be found "manually". But as the problem gets complicated, we need to apply more sophisticated resolution algorithms. The easiest algorithms to code are greedy algorithms (see the section 4.2). These are very fast and guarantee a valid coloring in all cases, but rarely do they give the best solution. In fact, sometimes they give the worst possible.

So we will also see other methods, either with guaranteed optimum result, but with slow, even unusable run time in practice on large graphs, or with stochastic ones, with an unguaranteed but still generally rather good result.

DOI: 10.1201/9781003477785-2

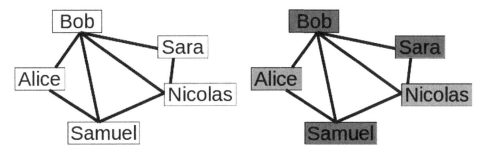

FIGURE 2.1   The edges of the graph represent incompatibilities. A three-coloring is possible; it will take three cars, one of which is just for Bob.

1	2	3	4
5	6	7	8
9	10	11	12
13	14	15	16

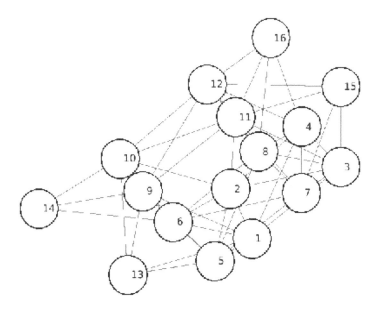

FIGURE 2.2   Sudoku – number of the boxes and corresponding graph.

2.2 SUDOKU

For another illustration of the use of graph coloring, consider a small Sudoku, 4 x 4. Figure 2.2 shows the numbering of nodes/boxes and the resulting graph. An edge indicates that the end nodes cannot contain the same number. For example, nodes 1, 2, 3, and 4 have to contain only different numbers. The same applies to nodes 1, 2, 5, 6, etc.

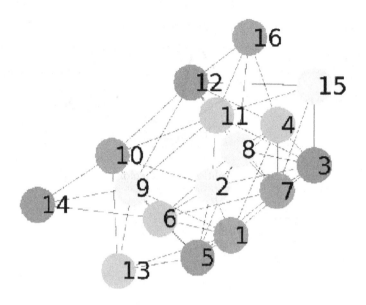

1	2	3	4
3	4	1	2
2	1	4	3
4	3	2	1

FIGURE 2.3   Sudoku – coloring solution

The search for a minimal coloring by the simplest of the algorithms that we will see below (Greedy 0, in Section 4.2.1) proposes a four-color solution, which, coded from 1 to 4, solves the problem (see Figure 2.3).

2.3 MEETING ROOMS

We can first form a simple table with time on the x-axis, which shows the overlapping time of meetings. To construct the graph, the nodes are the meetings, and if two meetings overlap, an edge is drawn between them (Figure 2.4). The coloring of the graph indicates that three rooms are needed (Figure 2.5).

Note that this is a graph of intervals (graph of). For this type of graph, a solution can always be found in polynomial time even on a conventional computer (not quantum).

	8h	9h	10h	11h	12h	13h	14h
Meeting 1	x	x					
Meeting 2		x	x	x			
Meeting 3	x						
Meeting 4			x	x	x		
Meeting 5		x	x				
Meeting 6					x	x	x

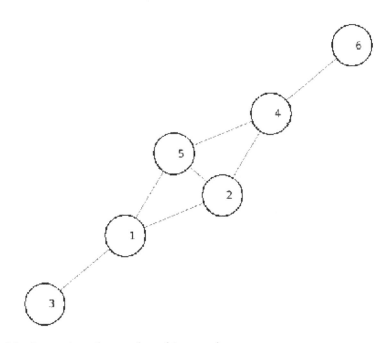

FIGURE 2.4   Meetings – time chart and resulting graph.

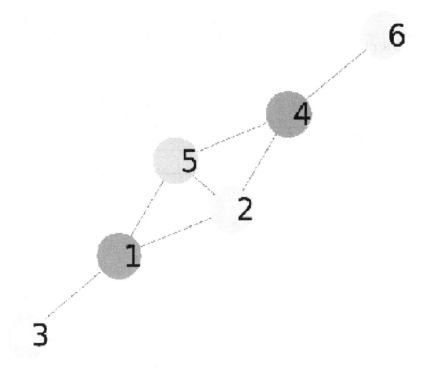

	8h	9h	10h	11h	12h	13h	14h
Meeting 1							
Meeting 2							
Meeting 3							
Meeting 4							
Meeting 5							
Meeting 6							

FIGURE 2.5   Meetings – graph resolved and assignment of the three rooms.

## CHAPTER 3

Encoding

To write resolution programs for a computer, it is necessary to know how to represent the entities of the problem to be solved according to codes usable by the machine (and the language used). Most often we rely on binary or more generally integer encodings.

3.1 DEFINITIONS

We have seen the definition of a valid coloring, which can be rewritten as follows:

A coloring is *valid* if two adjacent nodes are never of the same color.

We will need another notion of degree:

The *degree* of a coloring is the number of different colors used. It will often be denoted by K. Not to be confused with the degrees of the nodes.

From there, we can specify what the algorithms presented later will look for:

A coloring is *optimal* if it is valid for degree K and if there is no valid coloring of lower degree.

To simplify the presentations and the algorithms, we agree to represent the colors by integers. Hence another definition:

An integer coloring encoding is *permissible* if it contains each integer from 1 to K at least once and no other. From now on for simplicity we will often use the qualifying metonymic *permissible coloring* expression, especially in Appendix 8. Obviously it does not qualify the coloring itself, but the integer coding chosen from the infinity of possibilities (see below the equivalent colorings).

As the colors are arbitrary, under certain conditions, two colorings can actually be two representations of the same object:

Two colorings are *equivalent* if they are of the same degree and if all the nodes of the same color in one are also of the same color in the other.

For example, for the graph 2.1.1 the two colorings of Table 3.1 are equivalent.

Unless otherwise stated, all the colorings considered below will be assumed to be permissible and it should be noted that for any coloring, there is a minimal equivalent, the integer color code of which is the smallest possible (see Section 3.2.2). Thus, coloring (3, 1, 2, 5), which is not permissible, can be replaced by (3, 1, 2, 4). A Matlab® source code allowing this "compaction" is given in the appendix 8.7.2.

TABLE 3.1   Two Equivalent Colorings

Alice	green	red
Bob	red	blue
Nicolas	green	red
Samuel	blue	green
Sara	blue	green

3.2 ENCODING A GRAPH

Let us consider a graph with N nodes. We have already seen that it was practical and quite natural to number them from 1 to N. But you also have to indicate the edges. In the most general case, the graphs are oriented: each edge is represented by an arrow and can therefore be traversed only in the direction of the arrow. So there may be one edge from node i to node j, but not necessarily the other way around. However, we are only interested here in non-oriented graphs (bidirectional) and without loop (no edge of a node towards itself).

3.2.1 Binary Encoding

The graph can be represented by a G matrix of dimension $N \times N$, called adjacency matrix, such that $G(i,j) = 1$ if the $i - j$ edge exists (or the $j - i$ edge) and $G(i,j) = 0$ otherwise. Since the graph is not oriented, the matrix is symmetrical. For example, the five-node graph in Figure 3.1 has the following matrix:

$$G_{adj} = \begin{bmatrix} 0 & 1 & 0 & 1 & 0 \\ 1 & 0 & 1 & 0 & 1 \\ 0 & 1 & 0 & 1 & 1 \\ 1 & 0 & 1 & 0 & 1 \\ 0 & 1 & 1 & 1 & 0 \end{bmatrix}$$

From the point of view of storage in memory, and since the diagonal is zero, the symmetry allows to use only $N(N-1)/2$ bits, for example, those of the upper right triangle.

When the graph has few edges compared to the number of nodes, the matrix is sparse, containing practically only zeros. A condensed coding can then be interesting, in which the sequences of consecutive zeros are replaced by their length.

Another binary encoding consists of an *incidence matrix*, sometimes called a *vertex-edge matrix*. The edges are supposed to be numbered from 1 to A. The matrix is then $N \times A$. The (i,j) element is 1 if and only if the ith node and the jth node has an edge between them and 0 otherwise. The graph in Figure 3.1, assuming the edges numbered in a lexicographic order (i.e., $(1,2) \to 1, (1,4) \to 2, (2,3) \to 3, (3,5) \to 4, (4,3) \to 5, (4,5) \to 6)$), gives the following incidence matrix:

$$G_{incid} = \begin{bmatrix} 1 & 1 & 0 & 0 & 0 & 0 \\ 1 & 0 & 1 & 0 & 0 & 0 \\ 0 & 0 & 1 & 1 & 1 & 0 \\ 0 & 1 & 0 & 0 & 1 & 1 \\ 0 & 0 & 0 & 1 & 0 & 1 \end{bmatrix}$$

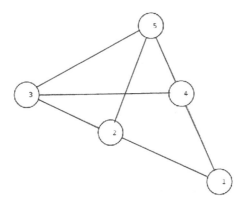

FIGURE 3.1   Five nodes, six edges graph.

For example, column 3 of the matrix tells us that nodes 2 and 3 are connected. An obvious property is that each column contains exactly two 1. Here, again, a condensed coding can be useful.

However, unlike the adjacency matrix, the incidence matrix alone does not fully describe the graph: it is also necessary to know the integer edge encodings. For example, in the example above, it should be noted that edge 2 is $(1, 4)$ and not $(1, 3)$.

This flaw can be avoided, but at the cost of a considerable increase in matrix size, using $N(N-1)/2$ columns, one per edge, existing or not (see Section 3.4.3).

More generally, all binary encodings, though easy to manipulate and practically indispensable for quantum calculations, are not economical in memory size. The entire coding and explicit list of edges, which we will see now, are much more concise.

3.2.2 Integer Encodings

We can consider the list of $N(N-1)/2$ elements of the matrix, which are all either 0 or 1 G_{adj} as being the binary representation of an integer. With our example: 1010101111, or 687 in base 10.

It is thus possible to define, for each N, "the set of integers that encode non-oriented N-order graphs without a loop". These sets have interesting non-trivial properties, starting with their size,[1] which increases very quickly with N: 1, 1, 4, 38, 728, 26704, 1866256, etc. On this set we can estimate the average efficiency of an optimal coloring search algorithm (see Comparisons 4.2.10).

3.2.3 Explicit List of Edges

The most intuitive is to indicate the edges as pairs of nodes. Thus, the graph in Figure 3.1 is represented by the following set of pairs:

$$\{(1, 2), (2, 3), (3, 4), (3, 5), (4, 5), (2, 5), (4, 1)\}$$

---

[1] Recursive formula $T_N = 2^{\frac{N(N-1)}{2}} - \sum_{k=1}^{N-1} \binom{N-1}{k-1} T_k 2^{\frac{(N-k)(N-k-1)}{2}}$

Since it is a set, the order of presentation of the pairs does not matter. Also since the graph is not oriented, neither is the order in each pair.

However, a more compact encoding can be used assuming that all node pairs are in a lexicographic order. Then, just give the ranks in this list of pairs corresponding to an existing edge. In our example, there are 10 pairs, six of which correspond to an existing edge

$$\{(\mathbf{1}, \mathbf{2}), (1, 3), (\mathbf{1}, \mathbf{4}), (1, 5), (\mathbf{2}, \mathbf{3}), (\mathbf{2}, \mathbf{4}), (\mathbf{2}, \mathbf{5}), (\mathbf{3}, \mathbf{4}), (\mathbf{3}, \mathbf{5}), (\mathbf{4}, \mathbf{5})\}$$

and the coding is then

$$\{1, 3, 5, 7, 8, 9, 10\}$$

(because the edge $(4, 1)$ is also the edge $(1, 4)$). This coding is more economical in memory space, but we will not use it here, as it is more difficult to interpret for a presentation.

3.3 ENCODING A COLORING

Of course, a coloring requires at most N codes. By convention, even if we speak of colors, we use integers of $I_N = \{1, ..., N\}$.

3.3.1 Natural Encodings

Since nodes are also supposed to be numbered from 1 to N, the encoding can simply be a list of integers taken from I_N (usually with repeats) with the convention that this list is in the order of the node numbers. For example the colorings of the graph in figure 3.3.1 are $(1, 2, 1, 2, 1)$ and $(1, 2, 2, 1, 2)$ for the one in figure 3.3.1. Note in passing that we can always impose that the color of node 1 is also 1. More generally, if there are K colors, we can always arrange things such that the coding includes only numbers at most equal to K. For example coloring $(1, 3, 5, 3)$ of a four-node graph uses three colors but is not permissible and can be replaced by the equivalent $(1, 2, 3, 2)$. That is what we assume now.

We will then see that purely algebraic manipulations on the coding of the graph and the coloring allow us to say whether the latter is valid or not.

However, for some coloring search algorithms (see Section 4.2.11.2), it is preferable to use a single integer encoding. Let $(c_1, ..., c_N)$ be the coding according to the method above. By definition, a c_n is at most equal to N. We can therefore consider, in the base $N + 1$, the integer code $code(C) = c_1 c_2 ... c_N$. Thus,

$$code(C) = \sum_{n=1}^{N} c_n (N + 1)^{n-1} \tag{3.1}$$

We thus define a total order relation on all possible colorings. Note that a lower bound $code_{inf}$ is for all c_n equal to 1 (all nodes of the same color, even if the coloring is necessarily not valid), and the maximum value $code_{max}$ is for $c = N - n + 1$ for the complete graph.[2] With this order relation, the smallest valid coloring is optimal.

This type of coding has strong formal links to what are called *pandigital* numbers (see the appendix 8.1).

---

[2] Recall that a bi-graphdirectional without loop is said to be "complete" if it has all possible $\frac{N(N-1)}{2}$ edges.

FIGURE 3.2   Two valid colorings whose binary encodings can be merged into a single matrix.

3.3.2 Binary Encoding

For some other methods, especially linear programming search (4.2.11) and the quantum approach (6), it is useful to have a purely binary representation, even though it is much less compact. Suppose that the nodes of the graph have K colors numbered 1, 2, …, K. We then define the matrix of coloring C of dimension $N \times K$ so that in each line n there is a single element equal to 1, in column k if the color of node n is k. All other elements of the row are null.

Thus, the binary coding of the coloring of Figure 3.3.1 is

$$C = \begin{bmatrix} 1 & 0 \\ 0 & 1 \\ 1 & 0 \\ 0 & 1 \\ 1 & 0 \end{bmatrix}$$

This code can be represented as a string of bits, by concatenating the lines. The example above gives 1001100110. Knowing N (or K) we can recreate the matrix.

REMARK

When it comes to just presenting in binary form the colorings found by an algorithm, the constraint "only one 1 per line" can be neglected. Consider, for example, the two colorings in Figure 3.2. They can be represented by the single matrix

$$C = \begin{bmatrix} 1 & 0 & 1 \\ 0 & 1 & 0 \\ 0 & 0 & 1 \\ 1 & 0 & 0 \end{bmatrix}$$

whose interpretation is that node 1 can have either color 1 or color 3.

3.4 CODING A COLORIZED GRAPH

So far we have coded graph G and coloring C separately. However, it can be interesting to have a coding that incorporates all the information: the structure of the graph and its coloring. We present the ideas here, though it is not used in the following sections.

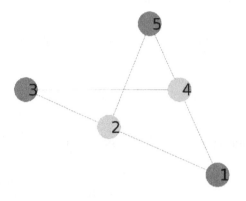

(a) A valid coloring (1, 2, 1, 2, 1).

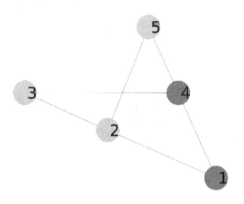

(b) A invalid coloring (1, 2, 2, 1, 2)

FIGURE 3.3   Two colorings of a five-node five-edge graph.

3.4.1 Use of the Diagonal

The first method starts from the adjacency matrix of the graph and completes it with the integer coding of C. It can be put on the diagonal or, with redundancy, on the values 1 of the matrix. Let's take the example of the five-node graph in Figure 3.3. With the coloring (1, 2, 2, 1, 2), the matrix becomes

$$
G_C = \begin{bmatrix}
1 & 1 & 0 & 1 & 0 \\
1 & 2 & 1 & 0 & 1 \\
0 & 1 & 2 & 1 & 0 \\
1 & 0 & 1 & 1 & 1 \\
0 & 1 & 0 & 1 & 2
\end{bmatrix}
$$

3.4.2 End Colors

A second method is to indicate the color of the end of each edge. This gives

$$G_C = \begin{bmatrix} 0 & 2 & 0 & 1 & 0 \\ 1 & 0 & 2 & 0 & 2 \\ 0 & 2 & 0 & 1 & 0 \\ 1 & 0 & 2 & 0 & 2 \\ 0 & 2 & 0 & 1 & 0 \end{bmatrix}$$

The matrix can then be read as

- edge of $1 \rightarrow 2$ is of color 2

- edge of $1 \rightarrow 4$ is of color 1

- edge of $2 \rightarrow 1$ is of color 1

- edge of $2 \rightarrow 3$ is of color 2

etc.

3.4.3 Binary Edge Coding

In this method, we build a string of bits by considering each edge successively (lexicographic order). If the edge is valid (ends of different colors), we add 1; otherwise we add 0.

The result does not in fact contain all the information of the (G, C) pair, because we can only deduce the number of edges, equal to the length of the string, but the code obtained can be useful during a search (see the section 4.1.3).

In our example, the code obtained is 100011, because

edge (1,2), valid, gives 1

edge (1.4), invalid, gives 0

etc.

## CHAPTER 4

Deterministic Resolutions

As mentioned earlier, we will present several types of search methods for optimal coloring, more precisely:

- deterministic but unguaranteed;

- guaranteed (thus necessarily deterministic);

- quantum

- heuristics and metaheuristics (inherently stochastic), but very succinctly.

> To avoid any confusion, let's clarify the meaning of some terms.
> A stochastic algorithm uses random numbers, unlike a deterministic algorithm. More precisely, it uses a random number generator (RNG). For practical reasons, the RNG is often algebraic, and therefore, strictly speaking, the algorithm is deterministic. Nevertheless, it is still called stochastic if, *in principle*, it can use a true RNG, for example, based on physical phenomena. See also Chapter 5.
> In the context of graph coloring, an algorithm is said to be "guaranteed" if it always produces a valid result in a finite amount of time.

All of the above have advantages and disadvantages and are not used in the same circumstances. For example, if the computing time is not a problem, you might as well choose a guaranteed method, even if very slow. Conversely, if the computing time budget is very small, a deterministic method without guarantee may be necessary. Heuristics and metaheuristics are compromises between these two extremes and are often very effective, but, as stated in the introduction, little will be said here.

Finally, let us recall that we are only interested here in generalist methods, valid for any type of graph because they do not exploit any structural features (flatness, tree structure, cycles, etc.), whether they are known in advance or detected by pre-processing. The only downside is that some algorithms still use degrees of nodes.

DOI: 10.1201/9781003477785-4

4.1 DIFFICULTY OF THE PROBLEMS AND LANDSCAPES

For a graph with N nodes, the total number of possible colorings is N^N. This is the size of the search space, which increases very quickly with that of the graph. In addition, the proportion of permissible colorings decreases with the size of the graph (see Appendix 8.3). Since not all permissible colorings are valid, we see that solving problems of increasing size means looking for needles more and more scattered in an increasingly large haystack.

We can consider the problem of finding a minimum coloring as a problem of optimization. To do this, we only need to be able to assign a value to each point of the search space—each possible coloring—in such a way that the desired coloring has the smallest value possible. We can even simply define a relationship of order between the colors and be able to say that such coloring is "inferior" to others. This is done, for example, for the sequential search of Section 4.2.11.2.

If you assign values, you can visualize the "landscape" of the problem, at least 2D or 3D sections. Let us do it here, by calculating the value of a coloring C of an N nodes graph G as follows:

- Let $f_{1,G}(C)$ be the number of edges whose ends are the same color. The maximum value is $\frac{N(N-1)}{2}$.

- Let $f_2(C)$ be the number of different colors used. The maximum value is N.

- The value of the coloring is given by $f_G(C) = f_{1,G}(C) + \alpha f_2(C)$.

We want that for two colorings C_1 and C_2 we have:

- If $f_{1,G}(C_1) = 0$ and $f_{1,G}(C_2) > 0$ then $f_G(C_1) < f_G(C_2)$. In other words, a valid coloring is always better than a non-valid coloring, even if it uses more colors.

- If $f_{1,G}(C_1) = f_{1,G}(C_2)$, then it is the number of colors that counts.

It is enough to have $\alpha < \frac{1}{N-1}$, for example, $\alpha = \frac{1}{N}$.

4.1.1 Landscape in N Dimensions

Consider then the small graph of Figure 4.1 and build the landscape of our function f on the search space of the colorings. It is four-dimensional, therefore difficult to represent, but Figure 4.2 gives two 3D sections, for which two colors have been fixed. Although the search space already contains 256 points, we note that the landscape has some regularity. This is due to the fact that the "closer is better (in probability)" assertion is true even for an *a priori* combinatorial problems as here (see Clerc 2007 and the discussion on positively correlated problems in Clerc 2015).

The landscapes presented in Figure 4.2 are formed by linear piece-wise interpolations between the points of integer positions. They suggest that it is possible to code the colors with continuous real values and to make the optimizer work on a continuous function. This

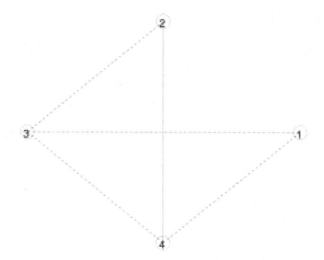

FIGURE 4.1   A four nodes five edges graph.

is outside the scope of this study, but by noting $c_i = C(i)$, A, the number of existing edges, and $\overline{A} = \frac{N(N-1)}{2} - A$, the number of non-existent edges, the following function can be used:

$$f_G(C) = \frac{1}{\overline{A}} \sum_{G(i,j)=0} \left(|c_i - c_j| > 1 \right) + \frac{\alpha}{A} \sum_{G(i,j)=1} \left(|c_i - c_j| < 1 \right) \qquad (4.1)$$

The first part counts the number of times two non-adjacent nodes have distant color values and will tend to minimize the number of colors. The second part counts the number of times two adjacent nodes have close color values and will tend to minimize such conflicts. The α coefficient gives more or less weight to this criterion. Once a minimum is proposed by the optimizer, it will be enough to replace the values c_i (possibly rounded to the nearest integer) by their ranks in an ascending order to find whole values coding an acceptable (permissible) coloring.

So, in our small example, with $\alpha = 2$, the standard particle swarm optimization algorithm SPSO 2007 (Particle Swarm Central) finds an optimal three-coloring in 40 fitness evaluations on average. For the cubic graph (8 nodes, 12 edges) presented in Section 4.2.1, it takes an average of 300 evaluations to find a bi-coloring. It certainly cannot compete with other sophisticated stochastic methods, but the important point is that it is possible to do much better than pure chance because the landscapes of coloring problems are not as chaotic as one might think.

Our example being really simplistic (planar graph), let us consider a more complicated one, like that of Figure 4.3 (7 nodes, 20 edges, and not planar).

However, as shown in the two sections of Figure 4.4, the landscape presents regularities usable by optimization algorithms, even if the presence of plateaus complicates the search.

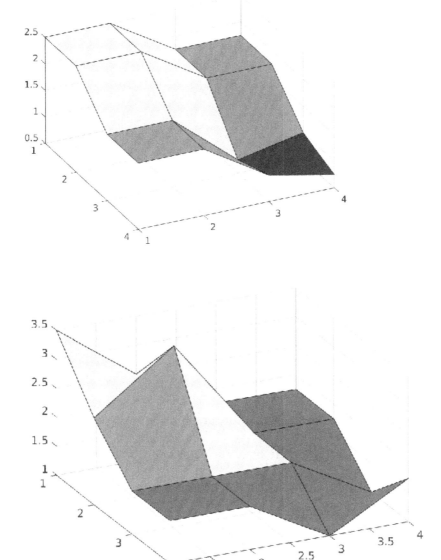

FIGURE 4.2 Landscape of coloring values for the graph of Figure 4.1. Two cross-sections. The first contains the minimum value 0.75 (valid three-coloring).

4.1.2 Landscape on Monocode

In fact, it is mainly the examination of landscapes on the search space of the coloring codes that can give the impression of chaos. Codes are defined here as the unique values assigned to each coloring, as in the 3.3.1.

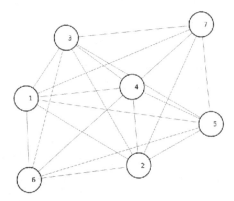

FIGURE 4.3   7 nodes 20 edges graph. Not planar.

Let us use a coloring evaluation simpler than the formula in equation 4.1. Let A_v be the number of valid edges, with different colored ends.

$$f_G(C) = \frac{A_v}{A} \qquad (4.2)$$

Figure 4.5 (with linear interpolations) shows an example of the cubic graph. We can see the minimum, but a classic iterative algorithm would have trouble finding it.

4.1.3 Landscape on Bicodes

A compromise is to define the coloring of the graph using two codes and thus to have a possible visualization in 3D for every N.

There are obviously several ways to build such a bicode. To illustrate the method let us consider two of them.

4.1.3.1 Binary Code, Supplement to 1

The first code of coloring is simply its binary code according to its edges, in the form of a bit string (see 3.3.2). The second code is its complement to 1. For example, 1001100110 and 0110011001 are transformed into integer (base 10), and to facilitate the representation, we add 1 and take the logarithms (base 2). The valuation is the same as for the monocode landscape.

It is completely artificial, and there is redundancy of information, but the landscape (with quadratic regressions) is more "sympathetic." Figure 4.6 shows an example for the same cubic graph.

The minimum should be easily found by a classic iterative optimizer, provided you know how to handle "colorings" with non-integer values.

4.1.3.2 Odd Bits, Even Bits

With the second method (which is equally artificial but without redundancy), we simply extract the bits of odd rank to form the first code, and the bits of even rank to form the

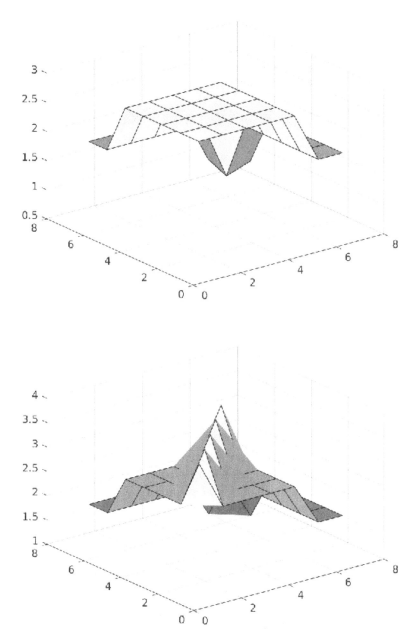

FIGURE 4.4   7 nodes 20 edges graph. Two 3D cross sections of the landscape of the colorings. Some regularities but plateaus complicate the search for the minimum.

second. For example 1001100110 gives $(10101, 01010)$. However, as the figure shows, the result is less convincing, especially since a possible algorithm should then find the two best minima to reconstruct an optimal coloring.

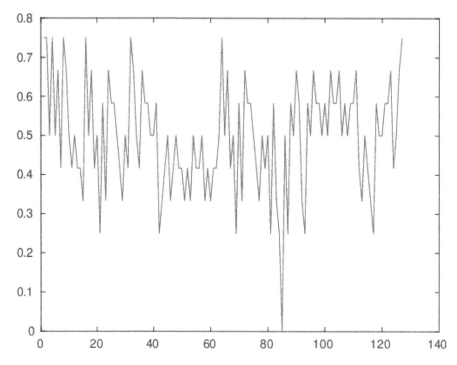

FIGURE 4.5   Cubic graph. Landscape on the two-color binary codes. The minimum is hard to find.

4.2 A FEW METHODS

The advantage of a deterministic algorithm[1] is that it is obviously useless to execute it several times, unlike stochastic methods (see Chapter 5). The disadvantage is that for unguaranteed deterministic methods, which are usually very fast and of low complexity, the result can be quite bad, if not *very* bad, on some problems.

The guaranteed deterministic algorithms are all, for the moment, of exponential complexity (on a classic computer) according to the number of nodes and edges.[2] However, in some cases, the complexity is polynomial, for example, if you know that the graph is three-colorizable (Wigderson 1982) or has a particular structure such as a tree, which we easily see it is always two-colorizable.

As seen before, we agree to represent the colors by integers of $\{1, 2, ..., N\}$. A node is said to be *vacant* if it is not colored. An important class of unguaranteed deterministic algorithms is known as *greedy*. Their general pseudo-code is:

---

[1] Sometimes called constructive, which is a bit restrictive, because some of the deterministic methods may have to destroy a partial coloring to try to rebuild a better one.

[2] Written in November 2023. This is related to the famous $P \overset{?}{=} NP$ problem. If you can prove the answer is yes, then, on the one hand, there is a classic (non-quantum) guaranteed deterministic algorithm of polynomial complexity, and on the other hand, you win the $1 million prize that the Clay Institute of Mathematics put into play in 2000.

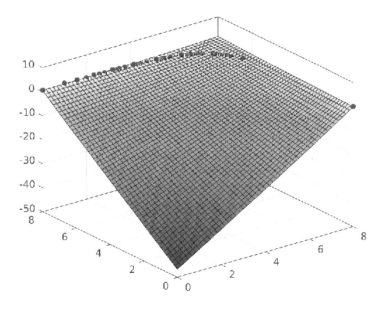

(a) Method 1 (binary code, supplement to 1)

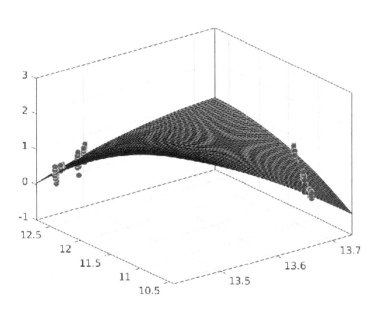

(b) Method 2 (odd bits, even bits)

FIGURE 4.6   Cubic graph. Landscapes on two-color bicodes.

> **GREEDY MODEL**
>
> As long as there is a vacant node
> - choose one
> - Coloring with a possible color
>   (i.e. not used by any of its neighbors)

Note that this model is a formalization of most strategies implemented by very young children in games like those of Chapter 1. As far as I could see, children who are a little older, say five or six years old, all use such algorithms, without of course naming them that way (see Section 4.2.9).

As we can see, when greedy algorithms assign a color to a node, they do not come back to it: no repentance in the artistic sense of the term. Thus, the complete coloring is done in N iterations. Starting from this general model, different variants focus on the method of choosing the next vacant node to be colored and the color to use. Here are some of them (several source codes are given in the appendix 8.7).

As we will see, small variations in reasonable intuitions can lead to algorithms with very different efficiencies.

4.2.1 Greedy 0

Select the vacant node of the lowest index and assign to it the smallest possible color that does not conflict with those of its already colored neighbors. According to the numbering order of the nodes, the algorithm finds an optimal coloring, or not! See Figure 4.7.

Even on a very simple "linear" graph with four nodes, the algorithm can fail, as shown in Figure 4.8 depicting the evolution of coloring. Two colors would be enough, but finally it offers three.

The theoretical interest is that there is always at least one numbering of nodes for which an optimal coloring will be found. The disadvantage, however, is that the total number of possible numbering is $N!$ (or, more precisely, $(N-1)!$ for you can always assign the color 1 to the first node), which increases very quickly with N and makes it unrealistic to try them all. A possible strategy, however, is to nevertheless try several other numbering options if it is suspected that the proposed coloring is not optimal with the given numbering of nodes.

But we can also build algorithms less sensitive to the numbering.

Another example is a simple cubic graph. It is classically called so, although in fact it is planar,[3] as shown in Figure 4.9.

It is bi-colorable but specially built to thwart certain algorithms: all nodes have the same degree, and the numbering is carefully chosen.

---

[3] But more generally a cubic graph is a graph in which all vertices have degree three, and it can be not planar. However, it always needs at most three colors.

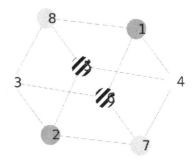

(a) **"Confusing"** numbering. The algorithm only offers a four-color solution.

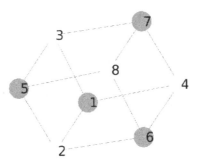

(b) **"Easy"** numbering. The algorithm finds an optimal bi-coloring

FIGURE 4.7   Greedy 0 on the cubic graph.

We can also imagine a "sober" algorithm, which, on the contrary, starts from the worst possible coloration, with N colors, and tries to reduce this number by considering the non-existent edges one by one. A source code is given in Appendix 8.7.5.4, and an exercise would be to prove that this algorithm is equivalent to Greedy 0.

4.2.2 Greedy 1

This is a slight variant of DSATUR (Brélaz 1979). It is like Greedy 0, but the nodes are previously ranked in a descending order of degree. In variant 1a, nodes are ranked in an ascending order of degree, but overall it is less effective (see Table 4.1 in the Comparisons section).

The algorithm is less sensitive to the numbering of nodes than Greedy 0. But it is still a bit sensitive because if several nodes have the same degree, it will select them successively according to the order given by this numbering. For each selected node, the assigned color is the smallest possible one.

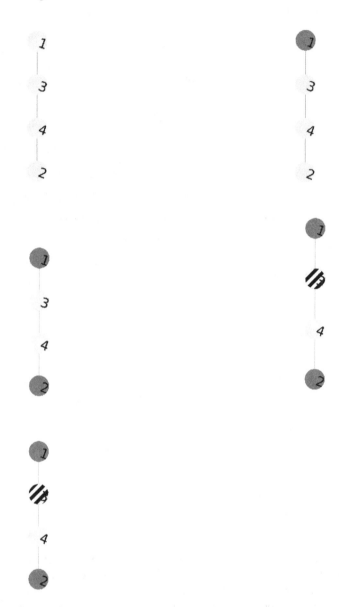

FIGURE 4.8   Greedy 0. Failure on a "badly" four-node numbered graph. The proposed solution has three colors, but two colors would suffice. The algorithm would find it with the numbering from top to bottom (1, 2, 3, 4).

In the cubic graph it is obviously not better than Greedy 0 (four colors), since the nodes all have the same degree.

4.2.3 Greedy 2

Choose the vacant node that has the largest number of vacant neighbors. The assigned color is the smallest possible one. Like Greedy 1 it is slightly sensitive to the numbering of nodes, if several have the same maximum number of vacant neighbors.

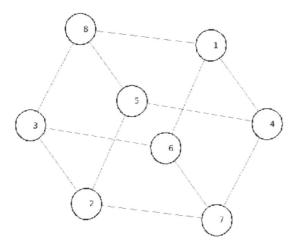

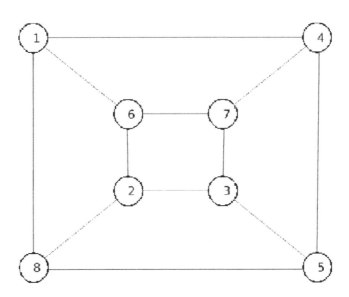

FIGURE 4.9   Cubic graph and planar representation.

But for the cubic graph it also offers only four colors, again because all the nodes have the same degree.

4.2.4 Greedy 3

We choose the vacant node to which we can assign a color of a lower number. There is still a risk of sensitivity to node numbering, if there are several possible candidates. Nevertheless, on the cubic graph it finds a bi-coloring (Figure 4.10).

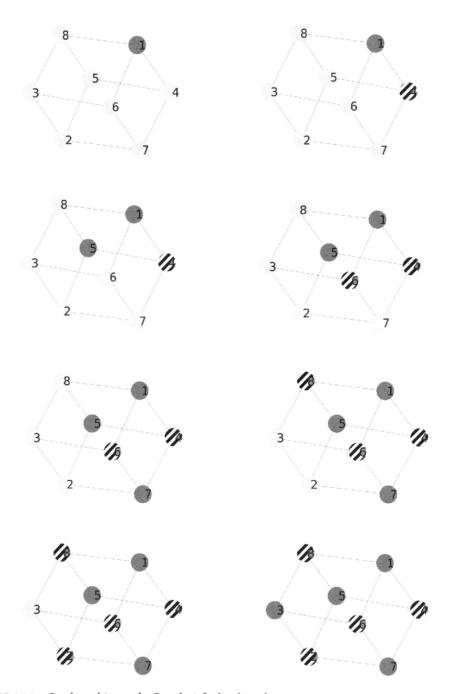

FIGURE 4.10   On the cubic graph, Greedy 3 finds a bi-coloring.

4.2.5 Greedy 4

When node i is colored, all pairs of (i, j) nodes are considered. Select the j that is vacant and whose path length from i to j in number of edges (the distance) is the smallest. Assign

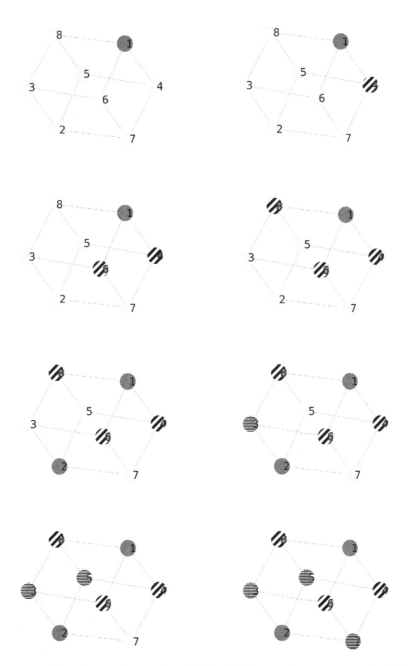

FIGURE 4.11 On the cubic graph Greedy 4 finds only three-coloring. Notice the difference with Greedy 3, at the fourth iteration.

the color of the lowest possible number. Same as for the previous ones: in the case of several equivalent choices, the numbering of the nodes has an influence. It is a little more complicated than the previous ones and involves more time-consuming calculation, and efficiency-wise it is not really any better (Figure 4.11).

TABLE 4.1 Average efficiency of Greedy algorithms. We note that Johnson, although significantly more complicated than several Greedys, is inferior and that RLF is slightly less effective than Greedy 2. Backtracking is very bad for small graphs, but unlike others, its efficiency increases with the number of nodes

Number of nodes ⇒	4	5	7	10
Greedy 0	0.992	0.953	0.839	0.627
Greedy 1	1	1	0.987	0.905
Greedy 2	1	1	0.994	0.936
Greedy 3	1	0.934	0.752	0.504
Greedy 4	0.994	0.965	0.842	0.624
Johnson	0.987	0.940	0.815	0.636
RLF	1	1	0.987	0.905
Greedy Eccentric	0.991	0.952	0.826	0.599
Backtracking	0.446	0.549	0.649	0.657
Sequential 1 (see 4.2.11.2)	0.934	0.952	0.880	

4.2.6 Greedy Eccentric

Let's define what is a "reasonable" greedy algorithm, even if the term sounds like a bit of an oxymoron:

- at each iteration it colors a single node;

- and with the smallest possible color (in the sense of integer coding) that does not conflict with those of its neighbors.

The first condition is not really binding since we can always go back to it. It simply makes it possible to affirm that for a graph of N nodes such an algorithm always builds a coloring in N iterations.

A greedy algorithm that would not be reasonable (eccentric? adventurous?) would, for example, assign color 3 to a node when 2 would be possible.

All the greedy methods seen so far are reasonable. Just out of curiosity it may be instructive to define and study one that is not.

For example, it is enough to start from Greedy 0, except that instead of assigning the smallest possible color, it assigns the largest one. This is a bit counterintuitive, but it appears that it seems only slightly less effective than Greedy 0 (see the comparison Table 4.1). Actually, a plausible conjecture is that over a large number of examples (but not closed under permutations, so that the No Free Lunch theorem does not apply), it is equivalent.

4.2.7 Johnson

This is another algorithm, quite old, a bit complicated (Johnson 1974), and moderately effective (see Table 4.1), but even if on the cubic graph, it finds a bi-coloring. It is presented here for historical purposes. It works by iterative reduction of the graph to its sub-graph not yet colored.

This does not appear explicitly, but it is sensitive to the numbering of nodes, like Greedy 0, and it can fail on very simple graphs, like the four-node linear (see figure 4.2.6). However,

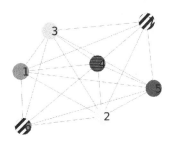

(a) Non-optimal solution on the 'linear' graph.   (b) 7 nodes, 20 edges, optimal solution (6 colors).

FIGURE 4.12   Johnson's algorithm, some examples.

it does find an optimal solution for the cubic graph and even for more complicated ones.

ALGORITHME DE JOHNSON (PSEUDO-CODE)

Color=0;
As long as there are uncolored nodes (vacant)
  Color=Color+1;
  G_current=G;
  As long as G_current is not empty
   G_current =
     G - (colored nodes and incidental edges)
   Find its vacant node of smallest degree
   Assign it Color
   Remove this node and its neighbors (Figure 4.12).

4.2.8 RLF

The *recursive large first* (RLF) method is significantly more sophisticated and designed for large graphs with relatively few edges (Leighton 1979). On a small graph, like the cubic, it finds only a four-coloring, but on a 10-node graph, it usually finds an optimal coloring.

First we define $v_c(i, j)$ as the number of neighbors common to i and j. Then we perform the iterative loop described in the box below. Subsets of nodes of the same color are progressively contracted into a single node.

RLF

1. Select a node i of maximum degree
2. Choose a node j not next to i such as
    $v_c(i,j)$ is maximum and contract (merge) j into i.
3. Repeat 2 until i is close to all the other nodes.
4. Remove node i and repeat step 1.
    All nodes contracted in i will have the same color (Figure 4.13).

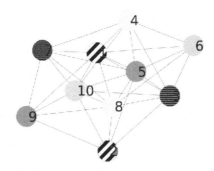

FIGURE 4.13   On this 10-node graph, RLF finds an optimal coloring of five colors.

4.2.9 Backtracking

So far we have only seen algorithms that gradually color the graph, with no possibility of going back. But, of course, we can build algorithms that admit a certain "right to error."

Here is one, whose principle is as follows:

- It is assumed that the graph is colorizable (valid) with K colors. Unless more precise information on the structure of the graph is known (for example, the presence of at least one triangular sub-graph), we start with $K=2$;

- Consider the nodes one by one, in a certain order, and try to color them validly with a color at most equal to K;

- if not possible, return to the previous node and change its color, then resume;

- if you go back and you get to the first node, you start all over again with $K = K + 1$.

Again, this method can be seen as a formalization of a strategy used in a game such as the Colorigraph (5).

In its simplest version (see code 8.7.5.11) the order of the nodes is that of their numbering, but one can refine, for example, by classifying the nodes in a descending order. Even so, the algorithm is sensitive to numbering as soon as there are nodes of the same degree. It easily fails on a simple graph (see Figure 4.14), but, unlike others, its average efficiency increases with the size of the graph (cf. Table 4.1).

Another way to present the functioning of this algorithm is to consider it as a tree path of incomplete colorings. The root is the coloring with only the first node colored to 1. Some branches can be "pruned" during the search, which accelerates it, but does not change the level of efficiency. Nevertheless, even so, its computation time remains much higher than that of greedy algorithms.

Other more sophisticated backtracking algorithms exist, but not really general. For example, the one defined in Zhou et al. (2014) works well only if a maximum clique search pre-coloring is performed and, especially, requires user-defined parameters, in particular the time spent searching for cliques and the number of implicit constraints to be calculated, such as that of Figure 4.15.

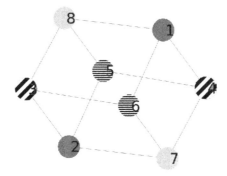

(a) Cubic graph. The solution found is 4 colors, while 2 are enough.

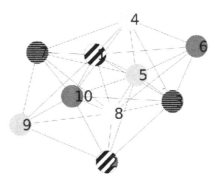

(b) Graph with 10 nodes. An optimal coloring with 5 colors is found, but not in the same order as with RLF.

FIGURE 4.14   Backtracking.

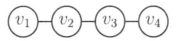

FIGURE 4.15   An implicit constraint is that for a two-coloring, vertices 1 and 4 cannot have the same color.

4.2.10 Comparisons

Let us define the efficiency of an algorithm on a given graph:

• it is 1 if an optimum coloring is found;

• it is 0 otherwise.

If two algorithms have the same average efficiency on all N-order graphs, we will say that they are N-equivalent.

For example, for $N = 4$, Greedy 1, Greedy 2, Greedy 3, and RLF are equivalent and even of perfect efficiency (equal to 1), but this is no longer the case for larger values of N. For $N = 10$ (with my small laptop, I can hardly go further), we find that the average efficiency of Greedy 0 is 0.627, while that of Greedy 3 is only 0.504. This result is also a bit unexpected because Greedy 3 is more sophisticated than Greedy 0. Table 4.1 gives more details.

We note that the most complicated algorithms are not necessarily the best (on average), RLF vs Greedy 2, for example.

However, as the number of possible graphs increases very rapidly with N, the indicated efficiencies are only estimates by random sampling. Naturally, in order for comparisons to be almost reliable, the same list of random graphs is used for all algorithms.

We can also, more classically, compare the results on some large given graphs. For example the problem wap05a.col of the DIMACS (2022) has 905 nodes and 43081 edges and its chromatic number is 50. Here are the results provided by some algorithms in Table 4.1:

- Greedy 0 finds coloring with 64 colors, Johnson and Backtracking with 63 colors;

- Greedy 1 and RLF find one with 51 colors;

- Greedy 2 finds an optimal coloring of 50 colors.

Some very sophisticated algorithms perform a pre-processing analysis of the graph, in particular they determine what we call its cliques (Gualandi and Malucelli 2012). We will not use that notion here, but a clique is a complete sub-graph; that is, any two nodes are always adjacent (see Figure 4.16). It is immediately clear that a clique of size K needs K colors to color it validly and, therefore, at least as much for the complete graph. A clique is maximal if there is no larger one (except possibly the complete graph).

However, such pre-processing does not necessarily yield better results, even in other examples.

4.2.11 Guaranteed Algorithms

If only to compare with algorithms that give a solution that is not necessarily optimal, it is useful to consider at least one completely guaranteed algorithm. The simplest is the exhaustive search that, of course, takes a considerable or even unacceptable time as soon as the graph is large.

As already mentioned, at the time of writing,[4] all algorithms guaranteeing an optimal solution in all cases have an exponential complexity in the size of the graph, at least on a conventional machine (Türing). But some still know how to take advantage of the structure of the graph to, quite often, be relatively fast.

---

[4] November 2023.

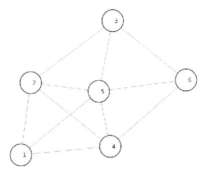

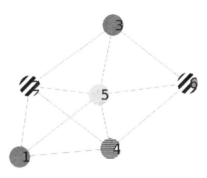

FIGURE 4.16   The sub-graph formed on the nodes (1, 2, 4, 5) form a maximal clique. It is then certain that it needs at least four colors, and therefore, the coloring presented (found by Greedy 0) is optimal.

4.2.11.1 Linear Programming

A method based on linear programming is given in the appendix 8.7.5.1. Binary coding of colorings is used. If we are looking for a K colors for a graph with N nodes, it can therefore be represented by a binary vector x of NK length. The basic principle is to try different values of K, starting from 2 (or a little more if we do a preliminary structural analysis) until we find one for which the system of equations and inequalities admits a solution. For each K:

- we build a matrix A_{eq} and a vector b_{eq} as we must have $A_{eq}x = b_{eq}$

- we build a matrix A and a vector b such that we must have $Ax \le b$

- we apply an algorithm to find a solution x, with the additional constraints that that all its components are 0 or 1.

Let us explain it with an example. Let us consider the four-node graph to the left of Figure 4.17. It is easy to manually find its optimal two-color coloring, shown to the right of the figure.

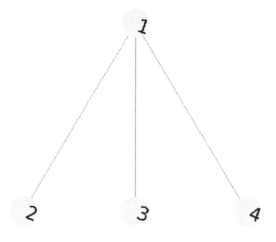

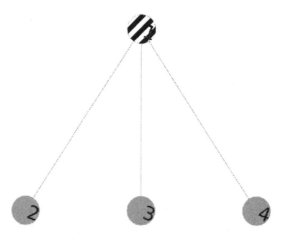

FIGURE 4.17 A four-node graph and its optimal coloring.

To find it by linear programming under constraints, consider its binary coding when trying with $K = 2$. It must be of the form

$$x = \begin{bmatrix} x_1 & x_2 \\ x_3 & x_4 \\ x_5 & x_6 \\ x_7 & x_8 \end{bmatrix}$$

Remember that the line of node i indicates its color by 1 in the corresponding column. By definition, then, there must be only one 1 in each line. A way to formalize it is to say that the sum on each line must be equal to 1. We rewrite x in the form of a "vertical" vector with eight components.

Hence the matrix

$$A_{eq} = \begin{bmatrix} 1 & 1 & 0 & 0 & 0 & 0 & 0 & 0 \\ 0 & 0 & 1 & 1 & 0 & 0 & 0 & 0 \\ 0 & 0 & 0 & 0 & 1 & 1 & 0 & 0 \\ 0 & 0 & 0 & 0 & 0 & 0 & 1 & 1 \end{bmatrix}$$

and the vector

$$b_{eq} = \begin{bmatrix} 1 \\ 1 \\ 1 \\ 1 \end{bmatrix}$$

so that we have $A_{eq}x = b$.

For inequalities we can use a direct algebraic calculation from the matrix G of the graph

$$G = \begin{bmatrix} 0 & 1 & 1 & 1 \\ 1 & 0 & 0 & 0 \\ 1 & 0 & 0 & 0 \\ 1 & 0 & 0 & 0 \end{bmatrix}$$

but it is more intuitive to gradually build the matrix A by considering the edges one by one. Here the vector b is

$$b = \begin{bmatrix} 1 \\ 1 \\ 1 \\ 1 \\ 1 \\ 1 \end{bmatrix}$$

For example, there is an edge between nodes 1 and 2, which cannot be the same color. So $x_1 + x_3 \leq 1$. Indeed, if they were of the same color, it would result in $x_1 + x_3 = 2$. So the first line of matrix A will be [10100000]. For the same reason $x_2 + x_4 \leq 1$ and the second line will be [01010000] etc.

Finally the matrix A is

$$A = \begin{bmatrix} 1 & 0 & 1 & 0 & 0 & 0 & 0 & 0 \\ 0 & 1 & 0 & 1 & 0 & 0 & 0 & 0 \\ 1 & 0 & 0 & 0 & 1 & 0 & 0 & 0 \\ 0 & 1 & 0 & 0 & 0 & 1 & 0 & 0 \\ 1 & 0 & 0 & 0 & 0 & 0 & 1 & 0 \\ 0 & 1 & 0 & 0 & 0 & 0 & 0 & 1 \end{bmatrix}$$

and the inequalities are of the form $Ax \leq b$.

The system to be solved is then

$$\begin{cases} A_{eq}x = b_{eq} \\ Ax \leq b \\ x_i \in \{0, 1\} \end{cases}$$

with N equalities and NK inequalities.

It will give as a possible solution (there are two equivalent) the coloring vector $x = (0, 1, 1, 0, 1, 0, 1, 0)$, which, when reshaped, becomes

$$\begin{bmatrix} 0 & 1 \\ 1 & 0 \\ 1 & 0 \\ 1 & 0 \end{bmatrix}$$

which, reinterpreted in full coding, gives $(1, 2, 2, 2)$. Naturally, if there is no solution, we will have to try another value of K. The strategy may be to increase from a sure lower bound or to start from a plausible value and increase by 1 in case of failure and decrease by 1 in case of success, or more economically, choose the successive values of K by dichotomy.

4.2.11.2 A Sequential Algorithm

Another way to build a guaranteed algorithm is to define an order relation on the colorings such that the "smallest" valid is optimal.

Coloring C_1 is said to be smaller than coloring C_2 in the following cases:

1. $\max(C_1) < \max(C_2)$

2. or $\max(C_1) = \max(C_2)$ and $code(C_1) < code(C_2)$

With this order relation, the search algorithm is relatively simple (see Box 4.2.11.2), especially if you code the colors from 0 to $N - 1$, because coding and decoding can then be done in base N. But it is necessary to indicate how to calculate the *next* of a code.

SEQUENTIAL SEARCH

```
K=2; % Or other lower bound if known
     % following a structural analysis
As long as 1=1 % A priori infinite, but we always find
         % a solution, at worst with K=N
% Cmin, permissible :
C=[0 <N-K+1 time>, 1, 2, ...,K-1];
% Can always be assumed, color 0 for the first node :
Cmax=[0, K-1 <N-K+1 time>];

As long as C is not equal to Cmax
if C is permissible AND valid
Propose C as a solution
END
C=next(C,K)
K=K+1 % No way with K, try one more color
```

An important point is to define how many times the loop will be executed.

The simplest method is to add 1 to the C code, then decode to generate the next coloring. In practice, we can generate the result directly, without having to code/decode.

TABLE 4.2 Sequence of colorings generated for four nodes, simply incrementing by 1. Colorings are indicated in Base 4. Unacceptable are marked with an asterisk. It would work for the complete graph with the solution (0 1 2 3), but for the linear graph, we find a coloring with three colors (0 0 1 2) because its code is smaller than the one with 2 colors (0 1 0 1)

C	C (following)
0 0 0 1	0 1 0 0
0 0 0 2*	**0 1 0 1**
0 0 0 3*	0 1 0 2
0 0 1 0	0 1 0 3*
0 0 1 1	0 1 1 0
0 0 1 2	0 1 1 1
0 0 1 3*	0 1 1 2
0 0 2 0*	0 1 1 3*
0 0 2 1	0 1 2 0
0 0 2 2*	0 1 2 1
0 0 2 3*	0 1 2 2
0 0 3 0*	0 1 2 3
0 0 3 1*	
0 0 3 2*	
0 0 3 3*	

Note that with this definition of the "next" one generates many unacceptable colorings (see example 4.2).

Consider the worst case for this algorithm, namely, the fully connected graph. The first coloring considered in base N, with two colors, is $C_{min} = (0, 0, ..., 0, 1)$. Then we increment.

One might think that it is enough to add 1. This would give the sequence of Table 4.2. It would also work for the fully connected four-node graph because only the last element (0 1 2 3) is valid and gives the solution.

But this is not the general case. With the linear graph in Table 4.2 one would find a valid three-color coloring before the optimal two-color solution.

Hence, the need to consider the sequences for different values of K. First the one for two colors then, if it fails (no valid coloring), the one for three colors, etc. (see pseudo-code 4.2.11.2 and source code in the appendix 8.7.5.12). This, for four-node, graph gives us the sequences of Table 4.3. Naturally, these sequences are only generated for the entire graph.

Or better, as already pointed out, one can, more economically, select the successive K by dichotomy.

For the complete five-node graph, it will thus test 318 colors, including 195 unacceptable ones, before finding the first valid and optimal coloring.

TABLE 4.3   A four-node graph. Sequences generated for successive values of K (number of colors tried). When it finds a solution the algorithm actually offers the equivalent "minimal" coloring. For example, (0 1 2 3) for the complete graph instead of (2 3 0 1). For the linear graph of Table 4.2 it almost immediately finds the solution (0 1 0 1)

$K = 2$	$K = 3$	$K = 4$
0 0 0 1	1 1 2 0	2 3 0 1
0 0 1 0	1 2 0 0	
0 0 1 1	1 2 0 1	
0 1 0 0	1 2 0 2	
0 1 0 1	1 2 1 0	
0 1 1 0	1 2 2 0	
0 1 1 1	2 0 0 1	
1 0 0 0	2 0 1 0	
1 0 0 1	2 0 1 1	
1 0 1 0	2 0 1 2	
1 0 1 1	2 0 2 1	
1 1 0 0	2 1 0 0	
1 1 0 1	2 1 0 1	
1 1 1 0	2 1 0 2	
	2 1 1 0	
	2 1 2 0	
	2 2 0 1	
	2 2 1 0	

Note that in practice, as graphs are rarely fully connected, the search often stops earlier. Table 4.4 shows the number of colors tested and the computing times before finding an optimal one for 10 random graphs of different dimensions. As can be expected, these numbers will grow exponentially with N, even if the search is relatively fast because in the vast majority of cases the algorithm eliminates the coloring as unacceptable (not permissible)[5] and therefore does not need to test its validity.

Figures 4.18 and 4.19 give two examples of graphs generated, with an optimal coloring obtained.

On the cubic graph the sequential algorithm finds an optimal solution after having tested only 85 coloring codes, all of which are permissible, which is exceptional, the average for eight nodes being about 54% (see Table 4.5).

NUMBER OF COLORINGS

For a graph with N nodes, one might think that the number of possible colorings is simply N^N since each of the N nodes can a priori be of any of the N colors. But in fact equivalent sequences count for only one coloring (see in the appendix 8.3). For example, sequences (3, 4, 4, 3) and (2, 3, 3, 2) represent the "same" coloring.

---

[5] Recall that a coloring $C = (c_1, ..., c_N)$ is permissible iff it contains at least once each integer of the $[\min_i (c_i), \max_i (c_i)]$ interval.

TABLE 4.4 Number of colorings checked (sequences) and computing time before finding an optimal coloring on ten randomly generated graphs for each number of nodes N. Unsurprisingly, the growth is exponential as a function of N, as can be seen from $N = 10$

N =>	Number of colorings checked					Computing time				
	4*	5	6	7*	10*	4	5	6	7	10
Trial 1	6	44	177	1946	789379	0.00209	0.00552	0.007562	0.0122	2.647
Trial 2	28	45	610	2545	863579	0.00224	0.00423	0.0117	0.0109	2.913
Trial 3	13	65	71	2084	768079	0.00210	0.00371	0.00532	0.00861	2.570
Trial 4	13	45	273	2195	38335	0.00178	0.00382	0.01424	0.00901	0.1304
Trial 5	17	14	173	1114	881996	0.00251	0.00412	0.00723	0.00591	2.955
Trial 6	13	63	143	2227	840218	0.00314	0.00456	0.008426	0.00917	2.795
Trial 7	6	145	153	1882	847084	0.03.19	0.00462	0.006115	0.00820	2.833
Trial 8	17	25	590	8079	884994	0.03.50	0.00289	0.01582	0.02609	2.932
Trial 9	13	137	253	210	696314	0.00298	0.00465	0.01079	0.00193	2.330
Trial 10	19	62	73	1200	789439	0.00275	0.00546	0.006453	0.00626	2.627

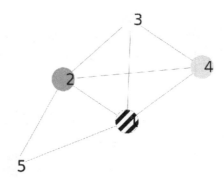

FIGURE 4.18   Sequential search. A five-node graph, optimal coloring (4 colors) found after 173 tested.

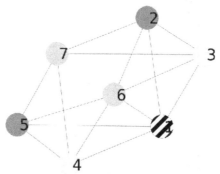

FIGURE 4.19   Sequential search. A seven-node graph, optimal coloring (4 colors) found after 2403 tested.

At first sight, sequential search is therefore not very effective, the greedy algorithms are much faster. But the essential difference is that the greedy algorithms do not always give an optimal coloring while here it is guaranteed.

Sometimes we can slightly decrease the number of codes to test. On the one hand, as with other algorithms, we can modify the numbering of nodes by considering more finely the structure of the graph (degrees of nodes, sub-graphs, etc.). More specifically, in some cases, it is possible to calculate a lower bound of the chromatic number greater than two. For example even a rudimentary structure analysis can highlight edge "triangles," in which case the algorithm can start with $K = 3$.

Or, as another example, it could be the classic Hofmann lower bound (Bilu 2006):

$$K = 1 - \frac{\lambda_{max}(G)}{\lambda_{min}(G)}$$

where $\lambda_{max}(G)$ is the largest eigenvalue of the adjacency matrix G and $\lambda_{min}(G)$ the smallest one (note that it is negative). There exist other bounds more sophisticated and often better (Edwards and Elphick 1983). In practice, however, these improvements remain marginal.

TABLE 4.5 The mean proportion of permissible coloring in those tested before finding a valid solution. Estimates based on 10,000 random graphs at most (for 10 nodes)

Number of nodes	Permissible/tested
4	78%
5	64%
6	60%
7	58%
8	54%

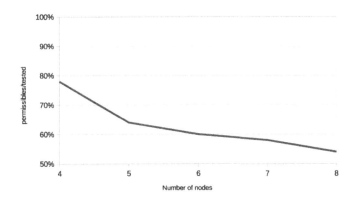

4.2.11.3 Improvements

A first improvement, obviously, is applicable when one reaches $K = N$. No need to check; an optimal coloring is $(1, 2, ..., N)$.

A second improvement is applicable at the very beginning, when trying $K = 2$. Indeed, all colorings are permissible, except the last $(1, 1, ..., 1)$. So no need to check the admissibility of successive colorings.

A third improvement is less obvious: it is a question of "skipping" certain non-permissible colorings. Figure 4.20 shows, for example, the sizes of these unproductive sequences for the seven-node graph in Figure 4.19. They represent a total of nearly 45% of the tested colorings, and according to the pseudo-code 4.2.11.2, the algorithm is forced to consider all these colors.[6]

But for each value of K it is relatively easy to find at least the largest of these sequences (cf. the appendix 8.4), and therefore, avoid checking the admissibility of the colorings it contains. By applying the method indicated in the appendix, only 2318 colors are tested instead of 2403, a gain of 3.5%. This is quite small, and moreover, this gain is generally decreasing with the size of the graph because, for each K, we skip only one sequence. So it would be interesting to find a way to do better and skip several sequences.

In addition, as shown in the histogram of Figure 4.20, the overwhelming majority of potential jumps (non-permissible coloring intervals) is null or very small. The most interesting intervals to jump are, on the contrary, very rare.

---

[6] By the way, most softwares (Matlab® here) are limited to the symbols taken in $\{0, 1, ..., 9\} \cup \{A, B, ...Z\}$ and therefore refuse a base greater than 36. For graphs larger than this size, it is sometimes necessary to write a specific program.

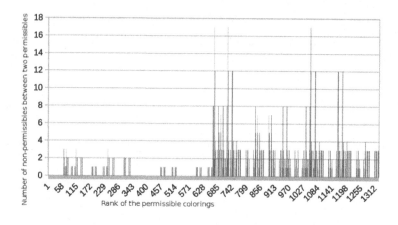

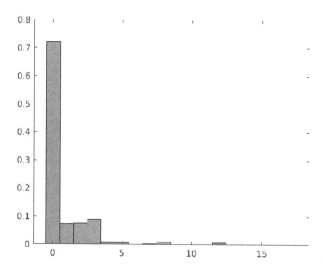

FIGURE 4.20   Sequential method on the seven-node graph in Figure 4.19. Number of non-permissible colorings between two permissible ones. They represent more than 45% of the graphs examined. Nevertheless, as shown in the histogram of the distribution of jump sizes, most are null or very small. It is the few large intervals that explain that percentage.

A fourth improvement is based on an empirical finding: knowing that the maximum number of edges is $m = N(N-1)/2$, the closer the graph is to this maximum, the more the likelihood of the chromatic number being close to N. In practice it is then often more efficient to start on a K value large enough and, depending on the result, opt for a decrease or an increase.

Let us take an extreme case to illustrate, the graph in Figure 4.21. It has 7 nodes and 20 edges. So it is almost complete ($m = 21$). The sequential algorithm based on $K = 2$ and even with the jump technique we have just seen tests 28,541 colorings before finding an optimal coloring with six colors.

But if we start from $K = 6$, for which we find a valid coloring, it is enough to then consider $K = 5$, for which no valid coloring is found. Thus, we convince ourselves that six is the

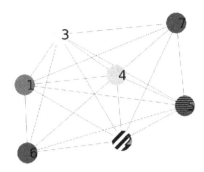

FIGURE 4.21   Sequential search. A 7-node 20-edge graph. Optimal six-color coloring found after 23,772 tested, starting from $K = 6$.

chromatic number. In doing so, the algorithm will have tested only 23,772 colorings, a gain of 16.7%.

We can also start from $K = 5$ and the algorithm, after failure, will try 6 successfully. More generally, if A is the number of edges of the graph and d_{max} the maximum degree of nodes, the following empirical formula for the initial value of K seems to give rather good results, especially for large graphs:

$$K_{initial} = \left\lfloor a\left(\frac{d_{max}}{A}\right)^b + 2 \right\rfloor \tag{4.3}$$

with $a = 0.037$ and $b = -1.4$.

For small graphs, it tends to underestimate the chromatic number. For example, for the graph in Figure 4.21 it gives only 2 instead of 6.

On the other hand, for the dsjc1000.9 graph (1000 nodes, $d_{max} = 924$, $A = 449449$) of the DIMACS library, it gives 215, which is interesting, because we know that the chromatic number is at most 222.[7] Thus, starting from 215 would most likely save considerable time calculation.

For a graph like wap05a (905 nodes, $d_{max} = 162$, $A = 21695$) the gain is even certain, since the formula gives 58 while the chromatic number is 50.

Note that in any case, even if a formula such as 4.3 is very coarse, it still allows a gain (relative to the fact of starting from 2) if $|K_{initial} - \chi| < |2 - \chi|$, where χ is the chromatic number of the graph.

4.2.11.4 Comparisons

We considered two guaranteed deterministic algorithms, by linear programming and sequential search. Let us compare them on a few examples, by applying, for sequential search, the improvements seen above. Computing times are only estimates for graphs of more than six nodes. For them we would have to process millions of random graphs, which my little computer cannot do in a reasonable time.

However, Table 4.6 shows that sequential searching is significantly more efficient than linear programming, even if the relative difference decreases with the graph size.

---

[7] Written in November 2023.

TABLE 4.6   Comparison of Linear programming and Sequential search. Mean computing time in seconds for randomly constructed graphs for each number of nodes. Sequential search is more efficient, but the gap decreases with the size of the graph

Number of nodes	Number of possible graphs	Number of graphs at random	Linear programming	Sequential search
4	38	100	0.0127	7.314×10^{-4}
5	728	2000	0.0174	7.273×10^{-4}
6	26 704	40 000	0.0182	0.00276
7	1 866 256	50 000	0.0702	0.1551

## CHAPTER 5

Stochastic Methods

The guaranteed methods induce long calculations, at least on conventional (non-quantum) machines. To overcome this disadvantage, we have seen fast deterministic but not guaranteed algorithms.

Another approach that is not guaranteed is to use randomness but in a controlled manner (Clerc 2015).

The search for optimal coloring can indeed be seen as a problem of minimizing a judiciously chosen objective function. In this case, heuristics or metaheuristics are used. The difference between the two is that heuristics are designed specifically for the type of problem under consideration, while metaheuristics have a more general field of application, even if it means making adaptations at times.

The distinction, however, is somewhat artificial: if the adaptations are sufficiently important, a metaheuristic is hardly anything other than a heuristic. In what follows, heuristics will simply mean "using randomness," as opposed to deterministic.

Of course, a really minimal coloring is not always found, but at least, even for large graphs, we often get an acceptable result in a reasonable time. It still remains to compare them to both greedy and non-greedy deterministic algorithms, which also do not always provide the best possible coloring.

Moreover, we note that almost all of these methods use algebraic generators of pseudo-random numbers, which are, in the final analysis, a list of perfectly defined numbers. So, strictly speaking, they are also deterministic algorithms, since the sequence of operations is reproducible from one execution to another (provided, of course, to use the same seed each time for the algebraic generator).

If we call them stochastic because the list is often so long that during an execution we do not risk having to use it cyclically. One can also have fun using relatively short and, therefore, precisely recycled lists, which is far from being ineffective (Clerc 2015, chapter 7), or an infinite list, but each element is directly accessible, like the decimals of π.

However, in principle, this type of algorithm can use true random numbers, coming from various physical phenomena, and their name is therefore not totally wrong.

Here, we can use the generic model used for deterministic algorithms (4.2), simply adding "at random" for the selection. The general pseudo-code then becomes:

DOI: 10.1201/9781003477785-5

STOCHASTIC MODEL

As long as there is a vacant node
- choose one at random
- coloring with a possible color
  (i.e. not used by any of its neighbors)

Naturally, "at random" actually means "controlled randomness," according to various criteria such as degrees of nodes or, more finely, degrees of saturation (the number of incidental edges whose other end is already colored).

Moreover, we see that, in the extreme, the level of randomness can be rendered zero and that, therefore, the deterministic model is only a special case of the stochastic model.

5.1 A FEW EXAMPLES

All heuristics and metaheuristics try to find the minimum of an objective function,[1] defined on a search space. In fact, more generally, it is enough that they are simply able to decide whether a position in the search space is "inferior" to another position. In other words, these methods just need an order relation on this space.

Here, for a given graph, the search space is the set of permissible colorings, on which we have already defined such order relation (see the section 3.3.1). Many others are also possible. For two colorings C_1 and C_2, we can rely on the notion of validity of colorings and on their degrees, denoted here by K_{C_1} and K_{C_2}, with the following two rules:

- if C_1 and C_2 are of the same status (valid or invalid), then the comparison is done on degrees. For example, $K_{C_1} < K_{C_2} \Rightarrow C_1 \prec C_2$ (read "C_1 smaller than C_2");

- if C_1 is valid and C_2 is not then $C_1 \prec C_2$.

But it's not necessarily the wisest approach. Let us simply remember that as soon as an order relation is defined on the coloring space, then any stochastic method is a priori usable. But which ones are the most effective?

As an example, and to make some comparisons, I have selected three representative methods described in R. M. R. Lewis's *A Guide to Graph Colouring: Algorithms and Applications* (Lewis 2021):

- an adaptation of the ant colony algorithm (ACO)

- an algorithm with returns (backtracking)

- a hybrid algorithm (hybrid)

---

[1] Except, of course, those that specifically deal with multi-objective problems and rather propose compromises between them.

TABLE 5.1 Comparisons between Greedy 2 and (meta)heuristics. Degree of best valid coloring found. For metaheuristics the number of evaluations that have been carried out is given in addition when it is high

Graphs ⇒ Nodes	Cubic 8	myciel3 11	myciel5 47	dsjc125.1 125	wap05a 905	dsjc1000.9 1000
(edges)	(12)	(20)	(236)	(736)	(43 081)	(449 449)
Possible colorings	16 777 216	2.8×10^{11}	3.8×10^{78}	1.3×10^{262}	∞	∞
Chromatic number	2	4	6	5	50	≤222
Algorithms						
Greedy 2	2	4	6	7	50	308
ACO	2	4	$6 (10^8)$	$5 (10^8)$	$50 (10^8)$	$252 (10^8)$
Backtracking	2	4	$6 (10^8)$	$5 (10^8)$	$50 (10^8)$	$302 (10^8)$
Hybrid	2	4	$6 (10^8)$	$5 (10^8)$	$50 (10^8)$	$283 (10^8)$

Their detailed descriptions are in the cited book, but here I am interested only in comparisons with Greedy 2, given in Table 5.1. We can see that the latter is far from ridiculous, even for a graph of 1000 nodes, and for a much lower search effort. This suggests that a good strategy might be to first apply Greedy 2 followed by a stochastic algorithm from the found solution.

5.2 A LAZY METHOD: THE BI-OBJECTIVE

An intuitive approach is to tell an algorithm our wishes:

1. minimize the number of colors;

2. minimize the number of invalid edges (same color at the ends).

and then to just say "Get on with this."

These two objectives are contradictory, hence the idea of using, precisely, a bi-objective algorithm that will produce a Pareto front of solutions that, for it, are equivalent (Figure 5.1).

This method is mentioned here because in practice all effective bi-objective algorithms are stochastic (Ulloa et al. 2020; Ahmadi et al. 2021). However, in fact, we will only consider one point on the front, for which objective 2 is zero. In the example, this corresponds to the value 5 for objective 1, which, in this case, is the chromatic number of the graph. Of course this is not always the case. On the one hand, the algorithm may not find a solution with zero for objective 2, and on the other hand, it may, and more frequently, like more conventional algorithms, offer only non-optimal solutions, with too many colors.

Let us give two examples, made with the bi-objective gamultiobj genetic algorithm integrated in Matlab®. A code, deliberately rudimentary to be more explicit, is given in Appendix 8.7.6.

For the myciel5 graph in the DIMACS library, which has 47 nodes and a chromatic number of 6, the Pareto front shows that the best solution found has 13 colors (Figure 5.2).

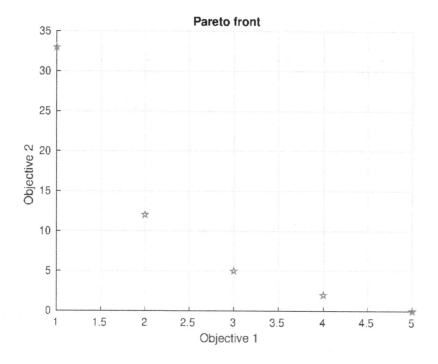

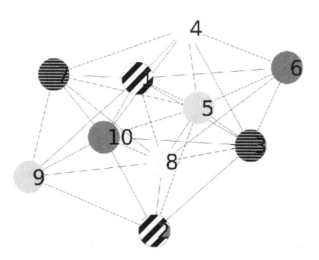

FIGURE 5.1   A 10-node graph treated by a bi-objective algorithm: Pareto front and optimal coloring.

It is certainly possible to refine this approach, for example by assigning different weights to the objectives, but, as already pointed out, this book does not really concern stochastic algorithms, and therefore, we will not further develop this presentation.

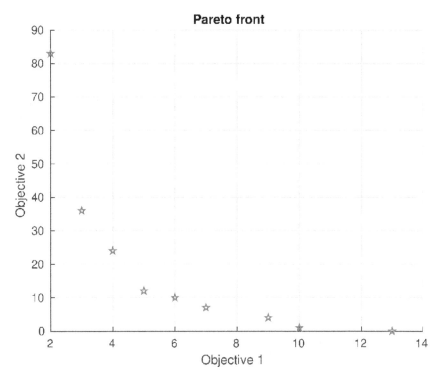

FIGURE 5.2   myciel5 (47 nodes) treated by the bi-objective algorithm: Pareto front. The best solution (objective 2 to zero) has 13 colors, while the chromatic number is 6.

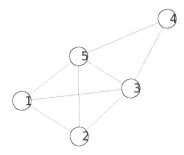

FIGURE 5.3   A five-node graph.

5.2.1 Random Search

You just designed your own stochastic algorithm. But do not forget that you have to compare it to the simplest one: the random search. The criterion is the number of colorings you have to evaluate before finding an optimal one. Here is how you can do that for two random search variants.

For a given number of colors K (of course we usually have to try different ones), we define a random coloring vector: N values chosen at random in $[1, 2, ..., K]$. Let us consider, for instance, the five-node graph of Figure 5.3 whose chromatic number is 4.

A random coloring could be $(2, 2, 1, 3, 1)$ and another one $(2, 1, 1, 3, 4)$. Both are not valid.

TABLE 5.2 Rudimentary random method on a five-node graph. Dichotomic search. Level of confidence 0.999. We had to sample 850 colorings to find an optimal one (four colors)

K	Number of runs	Result
2	107	failure
5	176	success
3	276	failure
4	291	success
Total	850	

A question is "What is the probability p to find an optimal valid coloring at the first attempt?" Or, more interestingly, "How many runs do we need to find at least one valid coloring?" The answers would be useful for possible comparisons with some other stochastic methods.

5.2.2 Rudimentary Method

For each node there are K possible values. So the total number of colorings is K^N. Let us define p as the probability of success (finding a valid coloring) at the first attempt (run).

For a given K, if there is a valid coloring, there are in fact at least $K!$ ones, by permuting the colors. This is the worst case for there may be more. So we have

$$p \geq \frac{K!}{K^N} \tag{5.1}$$

After r runs the probability of success is at least $1 - (1 - p)^r$. If we want success with a confidence level of ϵ (say 0.999), we just need to have

$$r \geq \frac{\ln(1 - \epsilon)}{\ln(1 - p)} \tag{5.2}$$

In our example, in order to find an optimal coloring, we can use dichotomy. After failure with $K = 2$ (see Table 5.2), we try $K = 5$ and find at least one valid coloring (with four colors and we can save it). So we try $K = 3$, unsuccessfully. Then, we try 4 and find a valid coloring (that we can save). So 4 is almost certainly the chromatic number (we are not completely sure; failure with $K = 3$ can be due to a lack of luck).

5.2.3 Smarter Method

Here we add a constraint: each color (value in $[1, 2, ..., K]$) must be used at least once. The number of colorings is $S(N, K)$, the number of surjections from a set of N elements to a set of K elements

$$S(N, K) = \sum_{i=1}^{K} (-1)^{K-i} \binom{K}{i} i^N \tag{5.3}$$

(see the appendix 8.3 for details). So the probability p is so that

$$p \geq \frac{K!}{S(N, K)} \tag{5.4}$$

TABLE 5.3   Smart random method on a five-node graph. Dichotomic search. Level of confidence 0.999. We had to sample 420 colorings to find an almost surely optimal one (4 colors)

K	Number of runs	Result
2	30	failure
4	240	success
3	150	failure
Total	420	

TABLE 5.4   Smart random method on myciel3.col (11 nodes). Dichotomic search. Level of confidence 0.999. The chromatic number is 6. We had to sample 3.96×10^6 colorings, but this is largely too much for, in the case of success, valid colorings are found many times

K	Number of runs	Result	Number of times a valid coloring is found
2	7060	failure	
10	376	success	226
6	1.24E+06	success	70819
4	1.01E+06	failure	
5	1.70E+06	failure	
Total	3.96E+06		

TABLE 5.5   Smart random method on myciel3.col. Dichotomic search. Level of confidence 0.1

K	Number of runs	Result	Number of times a valid coloring is found
2	108	failure	
10	6	success	4
6	18911	success	1111
4	15356	failure	
5	25996	failure	
Total	60377		

and for the number of runs we use the same formula 5.2 as for the rudimentary method.

In our example, again, we can use dichotomy. See Table 5.3.

For $K = 2$, no solution can be found, so we try $K = 4$ with success (no need to try $K = 5$ for in that case any coloring is, of course, valid). And then failure with $K = 3$.

Note that the estimated number of needed runs is often greatly overstated. It can be better seen on a bigger example, the myciel3.col graph from the DIMACS benchmark. It has 11 nodes. Optimal colorings are found many times as just one would be enough (see Table 5.4). Actually, we could even decrease the confidence level to 0.1, as we can see in Table 5.5. However, with an even lower level, the random search may not find any valid 10-color graph, and therefore, the dichotomy strategy would not work.

## CHAPTER 6

A Quantum Method

As noted above, it is generally impossible to obtain a really minimal coloring in a reasonable time for large graphs (except for special structures), even on super-computers,[1] as long as they are of the "Türing machine" type. It is no longer the same on a quantum machine. Certainly, one cannot be absolutely certain that the obtained coloring is optimal because quantum computing is inherently stochastic. But we can drastically increase the probability that this is the case by repeating the executions, while remaining within reasonable computation times.

However, there are two constraints:

- Having a powerful quantum computer. The progress is quite fast,[2] and we can hope to quickly reach enough qubits (quantum "bits") to solve the coloring of large graphs.

- Having suitable algorithms. In fact, it is the most uncertain part, because to develop such an algorithm, it is necessary, as Jean-Paul Sartre advocated in a completely different context, to "break the bones of the head," to reason differently from a conventional computer.

As an example of conceptual validation, an algorithm is presented below. If you do not have some notions of quantum calculation, you can easily ignore it and simply retain this conclusion: its complexity only increases polynomially with the size of the graph (instead of exponentially with classical algorithms). This can be understood intuitively because the power of a quantum computer increases exponentially with the number of usable qubits. Specifically, if n is this number, the number of simultaneously representable states is 2^n.

So, very roughly, since the number of ways to color an N-node graph with K colors is K^N, and if K is on the order of 2^k, we can understand that a quantum calculation with $n = kN$ qubits should allow to test them all at once. In practice, however, for large graphs, N is significantly greater than n, and the search must be split for different K values.

---

[1] Like Frontier (Oak Ridge National Laboratory, Tennessee), which, in 2022, could achieve 1102 PFlops/s (about 10^{18} operations/s).

[2] In July 2021 the Chinese Zuchongzhi processor had 66 qubits. In November 2021, the IBM Eagle computer had 127. In December, QuEra (Harvard University and MIT) announced 256 qubits. One year later IBM unveiled the Osprey, which boasts a massive 433 qubits.

DOI: 10.1201/9781003477785-6

Algorithms are given here in the form of quantum circuits. The constructions of these circuits are based on the mathematical method detailed in Appendix 8.5, but can be understood independently.

Otherwise, if you do not face the hidden traps and other pitfalls of these algorithms, you can always relax with another little game, given in the appendix 8.8 : Quantum Race!

6.1 BASICS OF QUANTUM COMPUTING

This section can obviously be skipped if you already know the quantum calculation. Moreover, this is only a very simplified presentation, just enough (and not quite orthodox!) to be able to understand the algorithms described afterwards.

Quantum bits or *qubits* are two-dimensional vectors of length 1. By convention, the vector $\begin{pmatrix} 1 \\ 0 \end{pmatrix}$ will be noted $|0>$ and the vector $\begin{pmatrix} 0 \\ 1 \end{pmatrix}$ will be noted $|1>$. These notations are called *kets,* and any qubit $\begin{pmatrix} \alpha \\ \beta \end{pmatrix}$ can be written as $\alpha|0> +\beta|1>$, in which α and β are complex numbers.

Note that $|\alpha|^2 + |\beta|^2 = 1$ is imposed, and therefore, $|\alpha|^2$ and $|\beta|^2$ are both in $[0, 1]$ and can be interpreted as probabilities. But probabilities of what?

On qubits, three types of operations are defined:

- *superposition*;

- *transformations*;

- *measure.*

An important concept in quantum computing is *entanglement.* This is not by itself an operation but a consequence of some operations on more than one qubit. See Section 6.1.1.

An algorithm is then often represented by a *circuit* which indicates, at each time step, which operations are applied to which qubits.

6.1.1 Measure

Let us start with the measurement, which is, very generally, the last operation of a circuit. One of the principles of quantum calculation is that, as long as a qubit is not measured, not only do we not know its value, which seems normal, but, more strangely, it does not actually have a really defined value, which is much less intuitive.

The result of the measurement is then to assign the value 0 with the $|\alpha|^2$ probability or the value 1 with the $|\beta|^2$ probability. So, after a measurement, a qubit becomes a classic bit. The usual method of defining an algorithm is therefore to manipulate qubits a little "blindly," without being able to know their state at every moment, but according to rules that, in the best case, will give the desired results finally, after measurements.

6.1.2 Poincaré-Bloch Sphere

Some transformations can be visualized by noticing that a qubit can also be written as

$$\cos\left(\frac{\theta}{2}\right)|0> +e^{i\phi}\sin\left(\frac{\theta}{2}\right)|1>$$

with $0 \le \theta \le \pi$ et $0 \le \phi \le 2\pi$. This defines a point of the unit sphere of $\mathbb{R}^3$ of cartesian coordinates

$$\begin{cases} x & = & \sin(\theta)\cos(\phi) \\ y & = & \sin(\theta)\sin(\phi) \\ z & = & \cos(\theta) \end{cases}$$

This representation will be especially useful for transformations that can be interpreted as rotations. Some websites offer interactive manipulations to better understand it (UTC 2023).

6.1.3 Transformations

A transformation can involve one or more qubits. We usually use the term "gate," in reference to the physical realization of a qubit by a photon that is transformed by actually crossing various devices. From a purely mathematical point of view, they are actually *operators*.

There are many, but let us comment only on those that will be used for the *NK* method (Section 6.2). All gates have an algebraic representation, using unitary matrices.[3]

6.1.3.1 H Gate

Initially, all qubits are $|0>$. So, to not bias any configuration, we make them «go through» an H gate (the name comes from Hadamard, which places them in the intermediate state, which would give 0 or 1 with probabilities equal to 1/2, if measured.

Such a gate therefore induces the transformation

$$|0> \to \frac{|0> +|1>}{\sqrt{2}} = \frac{1}{\sqrt{2}}\begin{pmatrix} 1 \\ 1 \end{pmatrix}$$

In matrix representation, we have

$$H = \frac{1}{\sqrt{2}}\begin{bmatrix} 1 & 1 \\ 1 & -1 \end{bmatrix} \tag{6.1}$$

We see that it also induces the transformation

$$|1> \to \frac{|0> -|1>}{\sqrt{2}} = \frac{1}{\sqrt{2}}\begin{pmatrix} 1 \\ -1 \end{pmatrix}$$

but we will not need it.

This is the analogue of the classic uniform random initialization often used in population-based algorithms.

---

[3] Recall that a unitary matrix is of norm 1 and that, therefore, applied to a vector, transforms it into another of the same norm.

6.1.3.2 X Gate

X stands for *exchange*. Sometimes called Pauli-X or NOT gate or *bit-flip*.

When a qubit "passes" through such a gate, it becomes, as it were, its opposite: $|0>$ becomes $|1>$ and $|1>$ becomes $|0>$. More generally α and β components are exchanged.

Its matrix is

$$X = \begin{bmatrix} 0 & 1 \\ 1 & 0 \end{bmatrix} \tag{6.2}$$

and to apply this gate is to perform the product

$$X \times \begin{pmatrix} \alpha \\ \beta \end{pmatrix} = \begin{pmatrix} \beta \\ \alpha \end{pmatrix}$$

This is the counterpart of Not in classical computing.

6.1.3.3 CX Gate

CX stands for *controlled exchange*.

It is also called CNOT because it is indeed a controlled "negation." This gate concerns two qubits. When the first one is $|1>$, an X gate is applied to the second one. But if it is $|0>$, the X gate on the second one is ignored.

So the matrix is now 4×4:

$$CX = \begin{bmatrix} 1 & 0 & 0 & 0 \\ 0 & 1 & 0 & 0 \\ 0 & 0 & 0 & 1 \\ 0 & 0 & 1 & 0 \end{bmatrix} \tag{6.3}$$

From an algebraic point of view, applying this gate consists of forming a four-element ψ vector by concatenating "vertically" those of the two qubits, then performing the product

$$CX \times \psi = \begin{bmatrix} 1 & 0 & 0 & 0 \\ 0 & 1 & 0 & 0 \\ 0 & 0 & 0 & 1 \\ 0 & 0 & 1 & 0 \end{bmatrix} \begin{pmatrix} \alpha_1 \\ \beta_1 \\ \alpha_2 \\ \beta_2 \end{pmatrix} = \begin{pmatrix} \alpha_1 \\ \beta_1 \\ \beta_2 \\ \alpha_2 \end{pmatrix}$$

In a circuit, this gate is often represented as

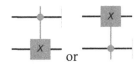

or, if you think especially of the "negation" aspect,

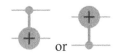

6.1.3.4 CnX Gate

This is a generalization of the CX gate, sometimes noted as MCX n (MC *multiple controlled*). Instead of a single control qubit, there are *n*, triggering the X gate if they are all $|1>$. Such a gate can obviously be replaced by a set of simpler ones, but when the quantum computer supports it directly, it allows to be more concise in the description of the algorithm. For example, in a circuit, a C3X gate can be schematized as

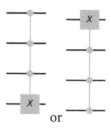

or

6.1.3.5 Z Gate

Its matrix is

$$Z = \begin{bmatrix} 1 & 0 \\ 0 & -1 \end{bmatrix} \tag{6.4}$$

In the Bloch sphere, it is equivalent to a rotation of angle π around the z-axis or, similarly, a symmetry in relation to this axis. It is also equivalent to the *HXH* succession, as the product of the three corresponding unit matrices easily shows.

6.1.3.6 CZ Gate

As for the X gate, we can define a controlled Z gate. Its matrix is then

$$CZ = \begin{bmatrix} 1 & 0 & 0 & 0 \\ 0 & 1 & 0 & 0 \\ 0 & 0 & 1 & 0 \\ 0 & 0 & 0 & -1 \end{bmatrix} \tag{6.5}$$

Its graphical representation in a circuit is symmetrical, as it does not matter which qubit controls the other.

In fact, of the four possible input states, namely, $|00>, |10>, |0,1>$, and $|11>$, only the last one is modified, transformed into $|11>$. To be convinced, consider, for example, the small circuit below.

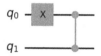

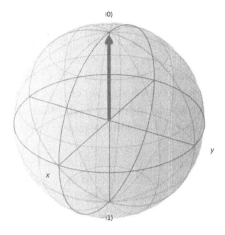

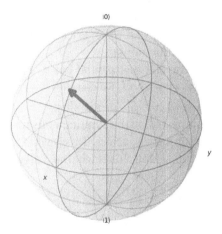

FIGURE 6.1. Effect of the $R_y\left(\frac{\pi}{4}\right)$ gate on a qubit in state $|0\rangle$.

Whether it is q_0 that controls q_1 or vice versa, the output state is always $|01 >$. With a CX gate we would have $|11 >$ in the first case and $|01>$ in the second. More formally, we can verify that in both cases the unit matrix of the CZ operator is the same.

6.1.3.7 Ry Gate

It is also a rotation gate, but more sophisticated, with the θ rotation angle being a parameter. Its matrix is

$$R_y(\theta) = \begin{bmatrix} \cos\left(\frac{\theta}{2}\right) & -\sin\left(\frac{\theta}{2}\right) \\ \sin\left(\frac{\theta}{2}\right) & \cos\left(\frac{\theta}{2}\right) \end{bmatrix} \qquad (6.6)$$

It will only be used during the initialization step of the NK algorithm. Figure 6.1 shows the effect of $R_y\left(\frac{\pi}{4}\right)$ on a qubit in the $|0 >$ state.

6.1.4 Superposition and Entanglement

These are undoubtedly the most delicate notions of quantum physics. There is no consensus on their interpretation.

For the *superposition* try to imagine how a photon can be *simultaneously* polarized right and left. And for the *entanglement* how two photons created together in a certain way (which is rightly called entangled) continue to influence each other, instantaneously, regardless of the distance that separates them (non-locality principle). Very counterintuitive (Einstein himself did not believe in it).

This was proved experimentally for the first time in 1982 (Aspect et al. 1982), then confirmed several times (Rauch et al. 2018; Moreau et al. 2019).

It should be noted that entanglement can be seen as extended superposition to several quantum objects.

But if we stick to the calculation itself, without seeking to know what it may well correspond to in the real world and thereby getting lost in quasi-philosophical considerations, there is no real difficulty because the predictions of these calculations are in good agreement with the physical measurements and, as far as we are concerned, with the results produced by quantum computers (or their simulations).

Thus, the superposition is in the fact that a qubit has two parameters (referred to above as α and β). If this qubit represents a photon, we can say that it is polarized on the right with an $|\alpha|^2$ probability and on the left with a $|\beta|^2$ probability.

As for entanglement, it is manifested by the simultaneous manipulation of several qubits, an algebraic operation based on unitary matrices. This is what CX gates do, for example, but we can of course process more qubits, and this is what we will do for the coloring of a graph.

6.2 NK METHOD

We are looking for a coloring of an N-node graph with a given number of K colors. The number of qubits to define a coloring matrix is NK (hence the name of this general method). In practice, to build a quantum circuit, the algebraic approach described in Section 8.5 is just a guideline that we do not need to follow strictly. The different steps are:

- Initialize, that is, generate a superposition of all possible colorings.
- Compare to graph. If for all edges the ends are of different colors, mark the coloring as valid, using an ancillary qubit set to $|1>$.
- Amplify valid coloring probabilities.
- Measure and propose solutions.

Each of these steps can be performed by a quantum sub-circuit. A complete Qiskit simulation code is given in the appendix 8.7.7.

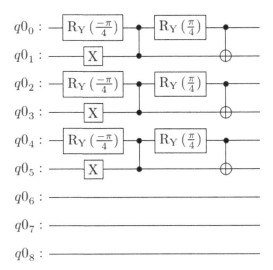

FIGURE 6.2   Generation of coloring matrices (W states).

6.2.1 Initialization

This step consists of generating a superposition of all possible colorings. The q_i qubits defining these colorings can be presented as an $N \times K$ matrix:

$$\mathbb{Q} = \begin{pmatrix} q_0 & q_1 & \cdots & q_{K-1} \\ q_K & \cdots & \cdots & \cdots \\ \cdots & \cdots & \cdots & \cdots \\ q_{(N-1)K} & \cdots & \cdots & q_{NK-1} \end{pmatrix}$$

Each of the N lines of K elements must have one and only one 1 , therefore in K possible ways. Thus, the total number of coloring matrices is K^N. In fact, this is done to realize W quantum states (named after one of its inventors Wolfgang Dür).

The algorithm first builds the initialization circuit that generates these W states, one for each node (Figure 6.2).

At its output, the state vector is then

$$\frac{\sqrt{2}}{4}|000010101\rangle + \frac{\sqrt{2}}{4}|000010110\rangle + \frac{\sqrt{2}}{4}|000011001\rangle + \frac{\sqrt{2}}{4}|000011010\rangle$$

$$+ \frac{\sqrt{2}}{4}|000100101\rangle + \frac{\sqrt{2}}{4}|000100110\rangle + \frac{\sqrt{2}}{4}|000101001\rangle + \frac{\sqrt{2}}{4}|000101010\rangle$$

Each binary sequence is read from right to left. The first NK bits represent a coloring matrix. For example, for the second state we have 010110, which gives

$$\begin{pmatrix} 0 & 1 \\ 1 & 0 \\ 1 & 0 \end{pmatrix}$$

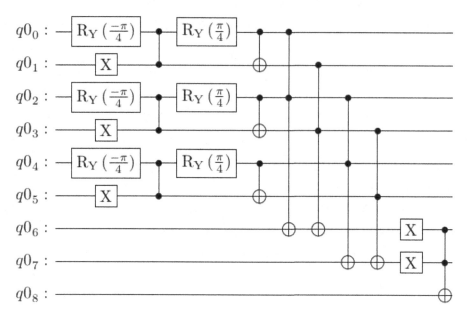

FIGURE 6.3 After initialization, marking of valid colorings.

meaning

- color 2 for node 1
- color 1 for node 2
- color 1 for node 3

which is actually an invalid coloring.

GREEDY MODEL

Be careful, these states are only visible in a simulation. On a quantum machine you can obtain one of them only after measurement. As they all have the same probability $\frac{1}{8}$, this would be useless at this stage. Hence, later, the need for an amplification of those corresponding to a valid coloring.

6.2.2 Mark Valid Colorings

An ancillary qubit is used for each arc of the graph. Using C2X and X gates, each of them is positioned at $|0>$ if both nodes have the same color or $|1>$ otherwise (valid arc) (Figure 6.3).

After this step the state vector becomes

$$\frac{\sqrt{2}}{4}|000010101\rangle + \frac{\sqrt{2}}{4}|000101010\rangle + \frac{\sqrt{2}}{4}|001010110\rangle + \frac{\sqrt{2}}{4}|001101001\rangle$$
$$+\frac{\sqrt{2}}{4}|010011010\rangle + \frac{\sqrt{2}}{4}|010100101\rangle + \frac{\sqrt{2}}{4}|111011001\rangle + \frac{\sqrt{2}}{4}|111100110\rangle$$

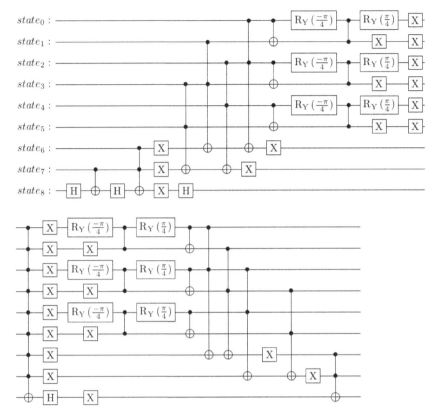

FIGURE 6.4   Detail of the amplification circuit of the states whose last qubits at $|1>$ (Grover's method).

Note that only the last two states correspond to valid colorings (the leftmost bit is set to 1).

6.2.3 Amplifying Probabilities

Naturally, increasing some probabilities necessarily leads to decreasing others, since their total must be 1. That is why a better name of this step is actually amplification diffusion.

Here, we use the classic Grover's method (Wikipedia 2023). The circuit is generated by the Qiskit code given in the appendix and presented in detail in Figure 6.4. We notice the quasi-symmetry due to the principle of the method, which applies a first circuit, then its inverse after a CnX gate, controlling the last qubit by all the others.

But in practice we can consider this circuit as a black box with all the qubits input, especially since this decomposition is not unique and depends on the tools used (machine and software).

Finally, the histogram of Figure 6.5 is obtained. For this very simple example, the probabilities of invalid colorings are virtually null (compared to the accuracy of the computer) and therefore do not appear. This will no longer be the case for the other examples. The two solutions are (omitting auxiliary qubits) 011001 and 100110. To be read from right to left

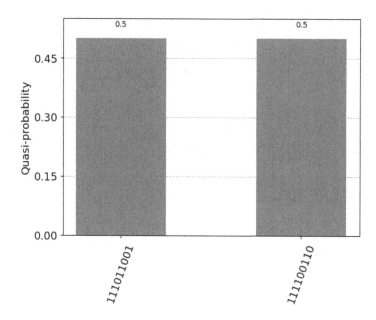

FIGURE 6.5   The two solutions for the three-node linear graph.

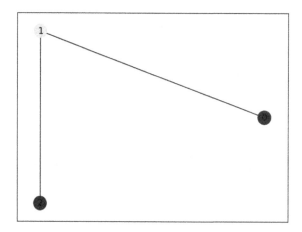

FIGURE 6.6   One of the two valid colorings of the three-node linear graph.

forming two-bit sequences, from where (10,01,10) and (01,10,01), that is to say the colorings (2,1,2) and (1,2,1). Figure 6.6 represents the second.

6.2.4 Examples

The following sub-sections discuss some examples where it becomes difficult to display the circuit and state vectors. They are nevertheless very simple, because, as we will see, the number of qubits required increases quadratically with the size of the graph.

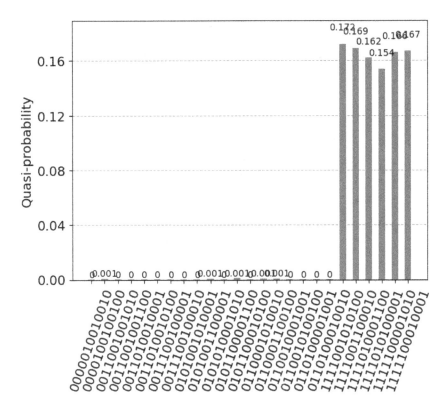

FIGURE 6.7   The six solutions for the triangular graph.

6.2.4.1 The Triangular Graph

If we are looking for a bicoloring, the answer of the algorithm is «No solution». With three colors, it simultaneously finds all six solutions,[4] as shown in Figure 6.7. Their probabilities are well amplified, but the others are no longer strictly null. So by being (very) unlucky, the measure might not give a valid solution.

6.2.4.2 Five Nodes, Five Arcs, Three Colors

Let's call the graph in question G5_3_3. A valid coloring found by the algorithm is given in Figure 6.8. Again the possible solutions (24) are found simultaneously (this is fundamentally different from conventional algorithm), but there are two disadvantages:

- The algorithm already requires 18 qubits.

- Even when amplified, the maximum probability becomes smaller and smaller with the number of possible colorings. So, sometimes it corresponds to an invalid coloring. In this case I had to run the algorithm twice for the maximum probability (of the order of 0.03) to correspond to a valid coloring.

---

[4] Assuming the algorithm runs on a real quantum machine, of course.

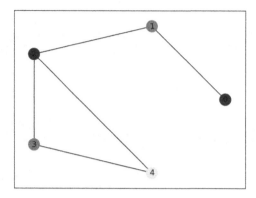

FIGURE 6.8   Graph G5_3_3, a valid three-coloring.

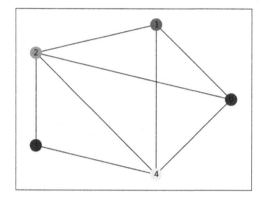

FIGURE 6.9   Graph G5_8_4, a valid four-coloring.

Incidentally, given the number of states, the histogram would be unreadable.

6.2.4.3 Five Nodes, Eight Arcs, Four Colors

Let's call this graph G5_8_4. Here you need 29 qubits. After two iterations of amplification, the algorithm finds all 48 solutions with a probability of approximately 0.01 for each, for example, 11111111110000001010000100001, which, by omitting the 6 auxiliary bits, is decoded from right to left as $(1000, 0100, 0010, 1000, 0001)$, giving the colors $(1, 2, 3, 1, 4)$ for the nodes $(0, 1, 2, 3, 4)$(Figure 6.9).

6.2.5 Complexities

For $K \geq 3$, determining whether any graph is K-colorizable is an NP-Complete problem. In practice this means that on a conventional computer (Türing machine), our simulations made with Qiskit require resources (memory, calculation time), growing exponentially with the size of the graph. So my 6 Gb laptop cannot even locally solve the three-coloring of a graph with four nodes and five edges (error message: *Insufficient memory*). For larger graphs, the code given in the appendix then uses a remote IBM machine.

We can estimate theoretical complexities, according to different measures. They are all polynomials of low degree. Other approaches are exponentially complex, even for the decision problem of colorability (Shimizu and Mori 2022). Still others suggest a polynomial complexity (Fabrikant and Hogg 2002), but only experimentally.

6.2.5.1 Number of Qubits

Let A be the number of arcs of the graph. We use

- NK qubits for coloring matrix superposition.

- As many auxiliary qubits as arcs. Maximum $\frac{N(N-1)}{2}$ for a complete graph.

- 1 ancillary qubit to mark valid colorings.

In all

$$Q = NK + A + 1 \tag{6.7}$$

As $K \leq N$ the complexity in the number of qubits is then

$$C_q = \frac{3N^2}{2} - \frac{N}{2} + 1 \tag{6.8}$$

that is, $O(N^2)$ with, moreover, a small coefficient (3/2). By the way, let's recall that Big-O notation can be misleading if we do not specify at least an order of magnitude for a possible coefficient.[5]

6.2.5.2 Number of Gates

Table 6.1 shows the distribution of gates by type.

For amplification the number of gates used must be multiplied by the number of iterations (see Section 6.2.5.4).

6.2.5.3 Size of the Circuit

A classic measure of the complexity of a quantum circuit is the product width×depth. The width is actually just the number of qubits used. The depth is the longest path in this circuit, between an input and an output. Its precise value depends on how the real circuit is realized. For example, some systems count one unit for a Z gate and others count three (because $Z = HXH$). Nevertheless, it is always a linear transformation (with small coefficients) on the number of gates, and therefore, this depth is of the order of $O(N^3)$.

It can also be requested in the Qiskit code. Table 6.2 and the attached figure, where complete graphs are considered, confirm a cubic complexity.

So, even with the worst possible strategy, which involves trying 2 colors, then 3, … then $N - 1$, the complexity of searching for an optimal coloring is in $O(N^4)$.

More cleverly, we can proceed by dichotomy on the number of colors K:

---

[5] For instance, in practice, an algorithm of difficulty cN^2 is not really better than an algorithm of difficulty N^3 if c is very big compared to N.

TABLE 6.1   Number of Gates

	Initialization	Marking	Amplification
H			4
X	N	A	$2(2A+1+N+NK)$
CX	$N(K-1)$		$2N(K-1)+1$
C_2X		AK	$2AK$
C_AX			2
CZ	$N(K-1)$		$2N(K-1)$
R_y	$2N(K-1)$		$4N(K-1)$
C_QX			1
Maximum			
$K=N$ and			
$A=N(N-1)/2$	$4N^2-3N$	$\frac{1}{2}(N^3-N^2)$	N^3+11N^2-8N+9

- try $K_1 = 2$

- if no solution, $K_2 = N - 1$.

- if no solution, N colors are required, one different for each node.

- if a solution, we know that the optimum is in $] K_1, K_2]$ and we try the integer K_3 closest to its middle

- etc.

At most we have to try k values with $2^{k-1} \leq N - 3 \leq 2^k$. The complexity then becomes

$$O\big((\ln(N-3)+1)N^3\big) \tag{6.9}$$

6.2.5.4 Number of Iterations

Grover's method requires a given number of iterations for maximum amplification. A probability of success that is neither too large nor too small follows a $\sin^2$ law. The theoretical formula is

$$i_{optim} = \frac{\pi}{4}\sqrt{\frac{number\ of\ states}{k}} \tag{6.10}$$

The parameter k is the number of solutions, which is *a priori* unknown. In many applications the number of states is 2^Q, where, remember, Q is the number of qubits. Then i_{optim} increases very quickly with the size of the problem.

Here, thanks to initialization (6.7) the number of states is reduced to K^N. Moreover, k is generally in the order of $O(N!)$. Thus, for the complete graph we have

$$i_{optim} = \frac{\pi}{4}\sqrt{\frac{N^N}{N!}}$$

By applying the Moivre–Stirling formula to $N!$, we see that i_{optim} increases less quickly but still as $\left(\frac{e^N}{N}\right)$, which remains fast. For example, for $N = 10$ we have to do 42 iterations, but for $N = 20$ we already get a value of the order of 5000.

TABLE 6.2 Complexity of the circuit for the *NK* method. It grows polynomially (cubic) with the size of the graph

Nodes	Width	Depth	Complexity
2	6	30	180
3	13	45	585
4	23	60	1380
5	36	75	2700
6	52	90	4680
7	71	105	7455
8	93	120	11160
9	118	135	15930
10	146	150	21900

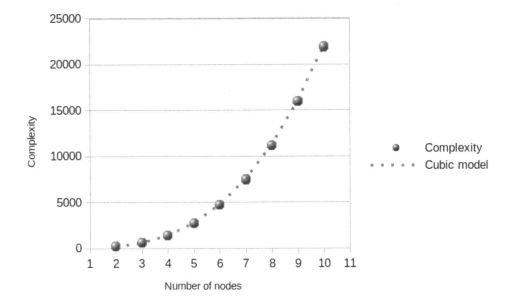

However, this is not so considerable if we note that the probability of finding a solution by random search is only $2,32 \times 10^{-8}$. On the other hand, it becomes prohibitive if the execution of the algorithm is only simulated on a conventional computer, because, in this case, the state vector must be kept in memory and it contains about 10^{26} elements.

## CHAPTER 7

Diplomatic Algorithms

The definition of a valid coloring is based on constraints like "not the same color for two adjacent nodes." For each pair of adjacent nodes, such a constraint has just two values: true or false. Thus, it is a binary and the algorithms that apply them are *intransigent*.

However, it is sometimes desirable to be more nuanced and to be able to indicate to what extent two adjacent nodes of the same color are acceptable.

This can be formalised by assigning *intolerance* levels to invalid arcs (ends of same color), for example, from 0 to 1. The value 1 means "totally unacceptable" and the value 0 means "perfectly acceptable" (and in that case the arc in question simply does not exist).

So we can calculate the *frustration* index of a coloring[1]

$$\frac{intolerance\ invalid\ arcs\ sum}{intolerances\ sum}$$

If the coloring is optimal, the index is 0, the frustration is 0, because there are no invalid arcs.

But if, as in some cases, one is forced to use fewer colors than the chromatic number, it is still necessary to reduce the number at best. An algorithm that seeks to do so, through judiciously chosen compromises, can be qualified as *diplomatic*.

Figure 7.1 gives an example with the triangular graph. Here, the tolerances are symmetrical, but this is not the general case (see below the example of carpooling).

Any intransigent coloring algorithm can be relaxed to become diplomatic. For example, the Greedy model seen in the section 4.2 becomes

---

DIPLOMATIC GREEDY MODEL

While there is a vacant node
-find the (vacant node, color) pairs
　that increases the frustration as few as possible
- if there is only one, then select it
　else apply criteria to select one
　　among these equivalent pairs

---

[1] By analogy with the concept of frustration in geometry and physics (e.g., for spins in antiferromagnetic interaction).

DOI: 10.1201/9781003477785-7

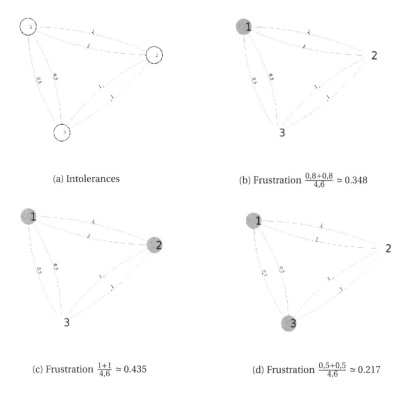

(a) Intolerances

(b) Frustration $\frac{0,8+0,8}{4,6} \simeq 0.348$

(c) Frustration $\frac{1+1}{4,6} \simeq 0.435$

(d) Frustration $\frac{0,5+0,5}{4,6} \simeq 0.217$

FIGURE 7.1   Triangular graph – possible intolerances and two-colorings. The best compromise is to assign the same color to nodes 1 and 3.

The clause after the "else" defines the possible variants. The choice of colors is made in a predefined list.

If the length of this list is greater than or equal to the chromatic number, there is a zero frustration solution, which may not be found, as for any Greedy method. More generally, even if this length is less than the chromatic number, the proposed solution, for which frustration will necessarily not be 0, will not always be the best.

For the Diplomat 0 algorithm, whose source code is given in Appendix 8.7.8, the "else" clause is simply "take the first encounter."

If all intolerances are equal, we obviously find Greedy 0, as shown in Figure 7.2: it takes four colors to get zero frustration, when two is enough.

But it is especially in the case where zero frustration is impossible that using a diplomatic algorithm is interesting. Let us take the example of carpooling from Section 2.1, but assuming there are only two cars available and assessing the levels of frustration if two people have to share the same car.

Figure 7.3 represents the estimated intolerances and a solution obtained by Diplomat 0.

The total frustration found is minimal and its value is $\dfrac{1+0,2+1+0,5}{10,7} \simeq 0,25$. Poor Bob (node 1) who, let us remember, hates everyone, is forced to share a car!

Very often, however, there are constraints that impose a lower bound smaller than the number of colors (resources). Let us take the example of carpooling again, but with seven

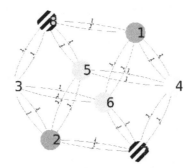

FIGURE 7.2   Cubic graph, treated by Diplomat 0, equivalent to Greedy 0 when all intolerances are equal.

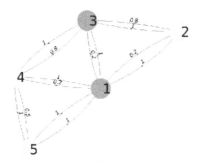

FIGURE 7.3   Carpooling of five people with two cars and intolerances.

people. Figure 7.4 shows us first a solution with two cars, which is unacceptable if each car can carry only four people. It takes three cars, Bob being alone in his car, and the frustration is zero.

In practice, therefore, we may have to launch a post-processing to check if the additional constraints are respected and, if they are not, to relaunch the algorithm with more colors/resources. This can, of course, be automated, but if the "budget" (usually a maximum number of iterations) is limited, there are two cases:

- there is no solution under this constraint;

- there is a solution, but the algorithm was unable to find it.

With a deterministic algorithm without guarantee, we can try others of the same type because, as we have seen, they are not all equivalent. But if the algorithm is deterministic guaranteed, it is hopeless. However, in general, it is rarely possible to use such an algorithm on big problems, so that leaves only the stochastic algorithms, relaunched multiple times and relying on luck. Alternatively, it is possible to use quantum algorithms, faster with better assured convergence, but on the condition of having a real quantum computer and not just a simulator.

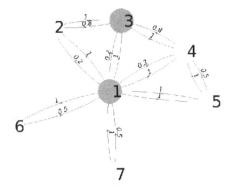

(a) With two cars, but one of them has to take five people.

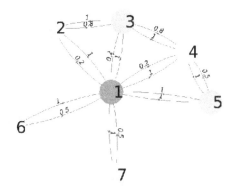

(b) With three cars. The best solution is still for Bob to be alone in his car and then the frustration is zero.

FIGURE 7.4   Carpooling of seven people and intolerances.

To conclude this brief overview on the concept of diplomatic algorithms, note that the scope is not restricted to problems of graph coloring, even with weighted arcs. All problems of satisfaction of constraints—for which there is already a vast corpus of methods—are concerned, precisely if there is no fully satisfactory solution. Then you get into artificial intelligence.

And that is a completely different story.

## CHAPTER 8

Appendix

This chapter presents some counts, a mathematical study underlying the quantum algorithm of chapter 6, and a number of source codes, so that the reader, if interested, can verify the given results and, possibly, conduct their own experiments and improvements.

Recall that "permissible (acceptable) coloring of degree K" is just a shortcut for "integer coding of coloring, containing only the integers from 1 to K (or from 0 to $K-1$) and each at least once." This is closely related to the notion of pandigital numbers. The number of permissible colors is an indication of the difficulty of the problem. Contrary to what one might think, it does not increase systematically with K.

8.1 PANDIGITAL NUMBERS

The general definition: a pandigital number is an integer whose writing includes all the digits (with or without 0) in a given base.

But in the context of sequential search for optimal coloring (see 4.2.11.2) let us modify it a little: we consider the numbers whose writing in base K includes exactly N digits, with possibly leading 0, and contains all numbers from 0 to $K-1$. Such a number could be called NK-pandigital, but we will simply call it pandigital.

For example, according to this definition, for $N = 7$ and $K = 6$ the number 0035412 is pandigital, but not 0035422 (1 is missing).

Thus, there is bijection between permissible colorings and pandigital numbers. For example, for a seven-node graph, the $(1, 1, 4, 6, 5, 2, 3)$ coloring is permissible and "identifiable" to the pandigital 0035412. This allows us some interesting calculations in particular concerning the forbidden (not permissible) coloring intervals (see Section 8.4).

8.2 EQUIVALENCIES

Consider a coloring of degree K. It defines a subset of N_1 nodes of color 1, a subset of N_2 nodes of color 2, ..., a subset of N_K nodes of color K. This can be represented by the list $L_K = (1, 2, ..., K)$ whose element of rank k is the common color of the nodes in the N_k subset.

DOI: 10.1201/9781003477785-8

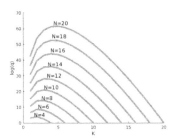

FIGURE 8.1 Number of equivalence classes (logarithmic scale).

Then the $K!$ permutations of L_K correspond to all colorings equivalent to each other. Because "colors" are arbitrary if, for example, we replace all 2 by 1 and simultaneously all 1 by 2, it does not really change the coloring.

Thus, each coloring of degree K belongs to an equivalence class of size K!. In summary, the set of all colorings of degree K can be partitioned into equivalence classes to the number of

$$q_{N,K} = \frac{A_{N,K}}{K!} = K^{N-K} \prod_{i=K+1}^{N} i$$

Among these, one and only one contains optimal colorings. A search algorithm will have to find it or, more precisely, one of its representatives.

As can be seen in Figure 8.1 for a given K $q_{N,K}$ increases rapidly with N, but for a given N it reaches a maximum for a low value of K, and then decreases fairly quickly.

8.3 PERMISSIBLE COLORINGS

Let us look at the total number of colors allowed for a graph with N nodes. A possible reflex is to say that each node can take up to N values and the number of sequences of N integers taken in $\{1, 2, ...N\}$ is then

$$A(N) = N^N$$

But every sequence is not necessarily a permissible coloring. For a coloring of degree K we have the constraint that each digit of $\{1, 2, ..., K\}$ must be used at least once.

In fact, this problem is that of calculating the number of $S(N, K)$ surjections of a set with N elements on a set with K elements. Indeed, if we consider a surjection of $\{1, ..., N\}$ on $\{1, ..., K\}$ it is clear that the constraint is satisfied.

The calculation formula is classic:

$$S(N, K) = \sum_{i=1}^{K} (-1)^{K-i} \binom{K}{i} i^N \qquad (8.1)$$

which can, for example, be demonstrated by recurrence using the fact that

$$K^N = \sum_{i=1}^{K} \binom{K}{i} S(N, i)$$

Finally, the total number of permissibles for a graph with N nodes is the sum of those for each degree K

$$S_{tot}(N) = \sum_{K=1}^{N} S(N,K) \tag{8.2}$$

8.4 NON-PERMISSIBLE INTERVALS

In sequential search (Section 4.2.11.2), let us define "jump" as the number of colorings between two permissible ones. What is the maximum jump?

We are interested here only in the case $2 < K < N$. Indeed, for $K = N$, finding an optimal coloring is trivial. Moreover, for $K = 2$, all jumps are zero.

It is then easy to see that the largest jump occurs for $K = N - 1$ and lies between the coloring $C_1 = (0, 0, N-2, N-3, ..., 1)$ and $C_2 = (0, 1, 0, 2, 3, ..., N-2)$.

The codes of these two colorings are

$$c_1 = \sum_{i=1}^{N-2} i(N-1)^{i-1}$$

$$= (N-1)^{N-2} - \frac{(N-1)^{N-2} - 1}{(N-2)^2}$$

and

$$c_2 = (N-1)^{N-2} + \sum_{i=2}^{N-2} i(N-1)^{N-2-i}$$

$$= \frac{1}{N-2}\left((N-1)^{N-1} + 2(N-1)^{N-3} - (N-1)^{N-2}\right) + \frac{(N-1)^{N-3} - N + 1}{(N-2)^2} - 1$$

For example, for $N = 7$, we therefore consider, in base 6, the numbers 0054321 and 0102345, that is, in base 10, 7465 and 8345. Their difference of 880 indicates that there are 879 non-permissibles between them.

The size of this maximum jump increases exponentially with N. For $N = 10$ it is already 1943079.

An obvious application is that if you do a sequential search for optimal coloring and you reach C_1 (because you have not found a solution for $K < N - 1$) then you can go directly to C_2.

But why stop when the road is so good? A similar calculation can be applied for $K = N - 2, N - 3, ..., 3$. Then we do the reverse, starting with $K = 3$, since the sequential search tests increasing values of K. Thus, generally speaking, if we arrive at $C_1 = (0, ..., K-1, K-2, ..., 1)$ jump directly to $C_2 = (0, ..., 1, 0, 2, 3, ..., K-1)$. However, for each value of K, only one sequence is skipped, and although this is the largest, the gain remains small, or more precisely, it is interesting only if the chromatic number is just slightly lower than N. Indeed, we must then test more K values, and therefore, we perform more jumps.

8.5 VALID COLORINGS

For a coloring, being permissible does not depend on the structure of the graph studied, only on its number of nodes, but of course what is actually sought is a valid coloring and having a minimum of colors.

It is possible to define the validity conditions of a coloring by a purely matrix approach. This is useful, for example, for linear programming resolution and also for a quantum approach. The C coloring is assumed to be coded by its binary matrix (which has, let us remember, one and only one 1 per row, the others being 0) and the G graph by its adjacency matrix.

We define

$$\Sigma = C \oplus C \tag{8.3}$$

$$G_K = \bigsqcup_K G \tag{8.4}$$

$$\Gamma = G_K \odot \Sigma \tag{8.5}$$

where $\oplus$ is the outer sum, $\bigsqcup_K$ the "horizontal" concatenation operator repeated K times, and $\odot$ the product element by element.

Then the indicator is

$$v_{C,G} = \max(\Gamma) \tag{8.6}$$

so that the validity of a C coloring for the G graph is given by the condition

$$v_{C,G} < 2 \tag{8.7}$$

which is equivalent to

$$v_{C,G} = 0 \lor v_{C,G} = 1 \tag{8.8}$$

which is easier to check by manipulating qubits. You can also use a modified outer sum so that Γ is binary. The condition is then simply

$$\Gamma = \mathbf{0} \tag{8.9}$$

where $\mathbf{0}$ is a null matrix.

```
The algorithm in Octave/Matlab® language
Sigma=outerSum(C);
% Variant:
% Sigma=max(0,Sigma-1);
GK=repmat(G, K,1);
```

```
Gamma=GK.*Sigma;
valid=max(Gamma(:))<2 ;
% Variants:
% valid=max(Gamma(:))<1; % valid=Gamma=zeros(size(C));
...
function Sigma=outerSum(C)
 % This a simplified version
 % The complete one has two matrices in input
 [~,K]=size(C);
    Sigma=[];
        for k=1:K
            Ck=meshgrid(C(:,k));
            Ck=Ck+Ck';
            Sigma=[Sigma Ck];
        end
 end
```

Proof

1. The only possible values in Σ are 0, 1, and 2;

2. The only possible values in Γ are 0, 1, and 2;

3. Let's consider $\Sigma(n, m)$. We have $m = (k-1)N + j$, with $j \in [1, 2, \cdots, N]$;

   (a) If $\Sigma(n, m) = 0$ neither node n nor node j has the color k;

4. If $\Sigma(n, m) = 1$ node n has the color k and node j another one or vice versa;

5. If $\Sigma(n, m) = 2$ nodes n and j have the color k (which is, of course, also true if $n = j$);

6. The value 2 is "eliminated" in Γ iff (if and only if) $\Gamma(n, m) = 0$, i.e., $G(n, j) = 0$, which means "no edge between n and j";

7. Therefore, if there is no 2 in Γ then the coloring C is valid.

Example for three-coloring (see Figure 8.2)

$$G = \begin{pmatrix} 0 & 0 & 1 & 1 \\ 0 & 0 & 1 & 1 \\ 1 & 1 & 0 & 1 \\ 1 & 1 & 1 & 0 \end{pmatrix}$$

$$C = \begin{pmatrix} 1 & 0 & 0 \\ 1 & 0 & 0 \\ 0 & 1 & 0 \\ 0 & 0 & 1 \end{pmatrix}$$

$$
\Sigma = \begin{pmatrix} 1 & 1 & 1 & 1 \\ 1 & 1 & 1 & 1 \\ 0 & 0 & 0 & 0 \\ 0 & 0 & 0 & 0 \end{pmatrix} + \begin{pmatrix} 1 & 1 & 0 & 0 \\ 1 & 1 & 0 & 0 \\ 1 & 1 & 0 & 0 \\ 1 & 1 & 0 & 0 \end{pmatrix}
$$

$$
\sqcup \begin{pmatrix} 0 & 0 & 0 & 0 \\ 0 & 0 & 0 & 0 \\ 1 & 1 & 1 & 1 \\ 0 & 0 & 0 & 0 \end{pmatrix} + \begin{pmatrix} 0 & 0 & 1 & 0 \\ 0 & 0 & 1 & 0 \\ 0 & 0 & 1 & 0 \\ 0 & 0 & 1 & 0 \end{pmatrix}
$$

$$
\sqcup \begin{pmatrix} 0 & 0 & 0 & 0 \\ 0 & 0 & 0 & 0 \\ 0 & 0 & 0 & 0 \\ 1 & 1 & 1 & 1 \end{pmatrix} + \begin{pmatrix} 0 & 0 & 0 & 1 \\ 0 & 0 & 0 & 1 \\ 0 & 0 & 0 & 1 \\ 0 & 0 & 0 & 1 \end{pmatrix}
$$

$$
= \begin{pmatrix} 2 & 2 & 1 & 1 & 0 & 0 & 1 & 0 & 0 & 0 & 0 & 1 \\ 2 & 2 & 1 & 1 & 0 & 0 & 1 & 0 & 0 & 0 & 0 & 1 \\ 1 & 1 & 0 & 0 & 1 & 1 & 2 & 1 & 0 & 0 & 0 & 1 \\ 1 & 1 & 0 & 0 & 0 & 0 & 1 & 0 & 1 & 1 & 1 & 2 \end{pmatrix}
$$

$$
G_K = \begin{pmatrix} 0 & 0 & 1 & 1 & 0 & 0 & 1 & 1 & 0 & 0 & 1 & 1 \\ 0 & 0 & 1 & 1 & 0 & 0 & 1 & 1 & 0 & 0 & 1 & 1 \\ 1 & 1 & 0 & 1 & 1 & 1 & 0 & 1 & 1 & 1 & 0 & 1 \\ 1 & 1 & 1 & 0 & 1 & 1 & 1 & 0 & 1 & 1 & 1 & 0 \end{pmatrix}
$$

$$
\Gamma = \begin{pmatrix} 0 & 0 & 1 & 1 & 0 & 0 & 1 & 0 & 0 & 0 & 0 & 1 \\ 0 & 0 & 1 & 1 & 0 & 0 & 1 & 0 & 0 & 0 & 0 & 1 \\ 1 & 1 & 0 & 0 & 1 & 1 & 0 & 1 & 0 & 0 & 0 & 1 \\ 1 & 1 & 0 & 0 & 0 & 0 & 1 & 0 & 1 & 1 & 1 & 0 \end{pmatrix}
$$

Note that values 2 in Σ are "eliminated" through the element-wise product by G_K. Let's look carefully at how Σ is built to find the indices of values 2. For each color k:

- Duplicate "horizontally" N times the corresponding column of matrix C in order to construct a square matrix Σ_k;

- add it to its transpose;

- if $k > 1$, concatenate "horizontally" the result to the previous one.

Thus, a value 2 is generated iff we have

$$
\Sigma_k(i, j) = \Sigma_k(j, i) = 1 \tag{8.10}
$$

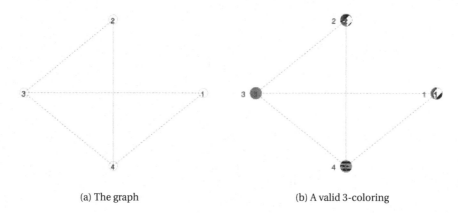

(a) The graph   (b) A valid 3-coloring

FIGURE 8.2   A four-node five-edge graph – an optimal coloring.

But $\Sigma_k (i, j) = \Sigma_k (i, 1)$ and $\Sigma_k (j, i) = \Sigma_k (j, 1)$, indicating that the same color k is given to nodes i and j. Thus, after a long detour, the rule is finally very simple:

There is a value 2 iff two adjacent nodes have the same color

This is useful among other things to build a quantum circuit. Here is an example of invalid coloring on our four-node five-edge graph (Figure 8.3):

$$C = \begin{pmatrix} 1 & 0 & 0 \\ 0 & 1 & 0 \\ 1 & 0 & 0 \\ 0 & 0 & 1 \end{pmatrix}$$

$$\Gamma = \begin{pmatrix} 0 & 0 & 2 & 1 & 0 & 0 & 0 & 0 & 0 & 0 & 0 & 1 \\ 0 & 0 & 1 & 0 & 0 & 0 & 1 & 1 & 0 & 0 & 0 & 1 \\ 2 & 1 & 0 & 1 & 0 & 1 & 0 & 0 & 0 & 0 & 0 & 1 \\ 1 & 0 & 1 & 0 & 0 & 1 & 0 & 0 & 1 & 1 & 1 & 0 \end{pmatrix}$$

CHROMATIC POLYNOMIAL

A conventional count is the one that uses the chromatic polynomial $P_G(K)$, which gives the number of valid colorings of degree less than or equal to K. For example, for the triangular graph, we have

$$P_G(K) = K(K-1)(K-2)$$

which shows that there is no valid coloring with one or two colors and that there are six with three colors. Here, since all three-colorings are valid, we can check that $A_{N,3} = P_G(3)$, where $A_{N,n}$ is the number of partial permutations of N elements n by n.

Of course, it is not always so simple. For example, for the cubic graph we have

$$P_G(K) = K^8 - 12K^7 + 66K^6 - 214K^5 + 441K^4 - 572K^3 + 423K^2 - 133K \quad (8.11)$$

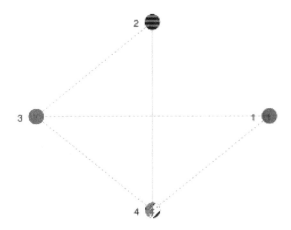

FIGURE 8.3   An invalid three-coloring. Adjacent nodes 1 and 3 have the same color.

In general the chromatic polynomial is therefore rather complicated. It can be built by recursion on operations of edge removal and graph contractions. The link with the chromatic number is the fact that the latter is the smallest integer for which the polynomial is positive. Once the polynomial is built, it is therefore easy to deduce the chromatic number, for example, to check whether a coloring algorithm finds an optimal solution or not (a source code usable for small graphs is given in Appendix 8.7.4)..

Thus, for the cubic graph, the polynomial is zero for $K = 0$ and $K = 1$ (which in any case is necessarily unacceptable for a connected graph) and is 1 for $K = 2$. It is therefore certain that it is two-colorable.

It would be interesting to have a formula for the number of valid colorings (number of colorings) $V(N, K)$. We have obviously

$$V(N, N) = N!$$

because if all the nodes are of different colors, we can swap them without risk.

More generally, assume that we have a valid coloring of K colors, not necessarily optimal. It defines a partition of N nodes in K subsets of the same color whose sizes are $(N_1, ..., N_K)$. Any color swapping between these sets defines another valid coloring, and all these generated colorings are equivalent. So we have

$$V(N, K) \geq K! \tag{8.12}$$

$K!$ is just a lower bound, as there may be other valid partitions, as shown in Figure 8.4.

8.6 DIFFICULTY RATING

To get an idea of the difficulty of the coloring problem, we can compare, for a given number of nodes N, the number of valid colorings to the number of permissible ones. But for this we

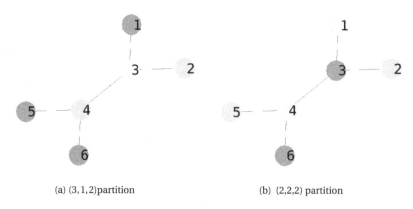

(a) (3, 1, 2)partition   (b) (2,2,2) partition

FIGURE 8.4 A graph can support multiple valid, non-equivalent colorings.

need the number of valid colorings or, failing that, an estimate, even rough, when we have only the lower bound. To date[1] there is apparently no exact formula published.

Let us start by estimating the average density of the connected N-graphs. For these graphs the minimum number of edges is $N - 1$, hence a density equal to $2/N$, and the maximum number $N(N - 1)/2$, giving a density of 1. We deduce that the mean density is

$$d(N) = \frac{1}{2} + \frac{1}{N} \tag{8.13}$$

Now, let us take inspiration from an experimental study (Gondran and Moalic 2018), suggesting an empirical inequality of the following form

$$V(N, K) < p + p^{\beta}$$

where p is the number of optimum colorings and $\beta \simeq 2$. However, it is difficult to apply because finding p becomes impossible in practice as soon as N increases too much.

8.7 SOURCE CODES

Codes are written in Octave/Matlab®. However, some instructions are specific to Matlab and do not work under Octave, mainly those for drawing graphs. A few small simple programs have been written in the language of Mathematica©, but are not given here.

8.7.1 Random Graph

Remember that we only consider non-oriented connected graphs without loop.

```
function edges=randConnect(N)
% Generate a connected graph with N nodes
Plot=false;
% Connect node 1 to another one j, at random
```

---

[1] November 2023.

```
connected(1)=1;
j=randi(N-1)+1;
connected(j)=1;
notConnectedList=2:N; % Remove 1
notConnectedList=notConnectedList(notConnectedList~=j); % Remove j
nbNotConnected=N-2;
connectedList=[1 j ];
nbConnected=2;
edges(1,1)=1; edges(1,2)=j;
nEdg=1; % Number of edges
% Connect the others
while nbNotConnected>0
 % fprintf('\n Connected \n'); disp(connectedList);
 %fprintf('\n Not connected \n'); disp(notConnectedList);

 % Select at random a non connected j
 j=notConnectedList(randi(nbNotConnected));
 %fprintf('\n j %i',j)
 % Randomly choose the number of links
 nbLinks=randi(nbConnected);

 % Randomly select nbLinks origins
 origins=randperm(nbConnected,nbLinks);
 %fprintf('\n  to connnect to \n'); disp(origins)

 % For each origin, create the link to j
 for n=1:nbLinks
  i =connectedList(origins(n));

  % Save the new edge
  nEdg=nEdg+1;
  edges(nEdg,1)=i; % i is connected ...
  edges(nEdg,2)=j; % ... to j
 end

 % j is now connected
 nbConnected=nbConnected+1;
 connectedList=[connectedList j];

 % Remove j from the list of unconnected
 notConnectedList=notConnectedList(notConnectedList~=j);
 nbNotConnected=nbNotConnected-1;
end
if Plot
 figure;
 G=graph(edges(:,1),edges(:,2));
```

```
h=plot(G,"-o",'MarkerSize',30,'NodeColor',[0.9 0.9 0.9]);
h.NodeFontSize = 20;
end
end
```

This construction method calls for some comments. To what extent, for a given number of N nodes, are the resulting graphs really random? Here, "random" is a shortcut for "generated by a uniform random process."

We have seen that any graph can be described by a binary sequence s. In practice, as we only consider non-oriented graphs without loop, we simply use the sequence given by the upper right triangle of the adjacency matrix. The length of this sequence is then $n = \frac{N(N-1)}{2}$.

Let m be the number of 1 of this sequence. If the latter is random, the probability that $m = k$ follows a Bernouilli distribution, in this case

$$\mathbb{P}(m = k) = \left(\begin{array}{c} m \\ k \end{array}\right) \frac{1}{2^n}$$

For example, let us simulate the process as if we wanted to generate 10^6 random graphs of $N = 30$ nodes purely randomly, without connectivity constraint, by generating sequences of $n = \frac{30 \times 29}{2} = 435$ bits and plot the histogram of the ratios

$$p = \frac{m}{n}$$

Then, m is the number of edges of the graph represented by the binary sequence. As expected, we obtain a Bernouilli distribution, symmetrical of mean $1/2$.

Now let us do the same using the above code of graph generation. In this case we have

$$m \geq N - 1$$

which implies

$$p \geq \frac{2}{N}$$

If the code indeed generates connected random graphs, the histogram obtained is expected to look like a Bernouilli distribution of mean $\left(\frac{2}{N} + 1\right)/2$, in our example 0.53333. And that is indeed what we find. The distribution shown in Figure 8.5 is "good looking," and its mean is 0.533295.

8.7.2 Equivalent Goloring

```
%{
  (Modified to more easily generate C++ code)
  Equivalent coloring, but so that it is as "small" as possible
  (according to the later coding of Cequiv by an unique integer)
  Example
  [5 5 4 3 3 2 2 2 2 4] => [1 1 2 3 3 4 4 4 4 2)
```

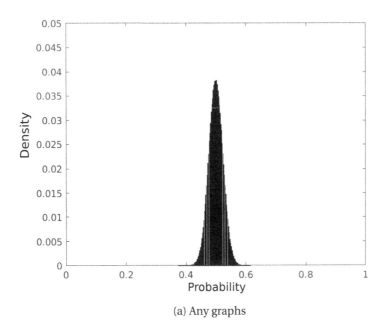

(a) Any graphs

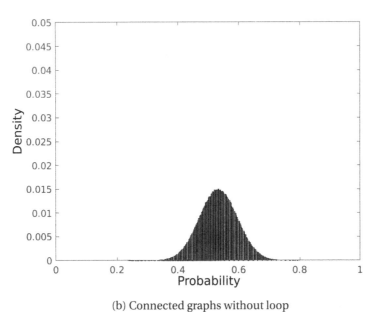

(b) Connected graphs without loop

FIGURE 8.5   Graphs of 30 nodes. Distribution of the number of edges, estimated on 10^6 random graphs. For connected graphs without loop, the mean is necessarily greater than 1/2.

```
%}
% Would be quicker by using unique(), but there is an issue when
% translating it into C++ with Matlab Coder.
Cequiv=zeros(1,N);
ind=find(C==C(1));
Cequiv(ind)=1;
```

```
k=1;
for n=2:N
    if Cequiv(n)>0 continue;end
        k=k+1;
        ind=find(C==C(n));
        Cequiv(ind)=k;
    end
end
```

8.7.3 Number of Connected Graphs

It is just the direct application of the classic formula. In practice, except on a very powerful computer, this code is unusable beyond a few dozen nodes. Note that the calculated number corresponds to structurally different graphs. For some algorithms, node numbering has an influence on efficiency. So the number of graphs they "see" as different is actually TN!.

```
% Octave/Matlab code
function T=nbConnectG(N)
%{
Number of undirected connected graphs with N nodes
See http://oeis.org/A001187)
1, 1, 4, 38, 728, 26704, 1866256, ...
Recursive code
%}
T=2^(N*(N-1)/2);
for k=1:N-1
c1=(N-k)*(N-k-1)/2;
u=nchoosek(N-1,k-1)*2^c1*nbConnectG(k);
T=T-u;
end
end
```

Considering all connected N graphs suggests an interesting thougt experiment. Of course, the numbers quickly become huge ($1, 57 \times 10^{57}$ for $N = 20$), but suppose that an user often has to look for an optimal coloring for graphs of size always at most equal to a certain N. As we have seen, it is possible to code each graph by a single integer.

So we can imagine that, once and for all, we establish a (gigantic!) sorted list containing, for each graph, its code and an optimal coloring found by a safe (and therefore long) method.

If we have to study a given graph, we have to be able to find it in the list. Consider, for example, the dichotomy method. The maximum number of trials t is such that

$$2^t \leq T(N) \leq 2^{t+1}$$

which implies

$$t \simeq \frac{\ln(T(N))}{\ln(2)}$$

But it appears that $\ln(T(N))$ follows a quadratic law (see Figure 8.6). Thus, once the list has been compiled—but unfortunately it is unrealistic as soon as N is large—the search for an optimal coloring is itself of quadratic complexity.

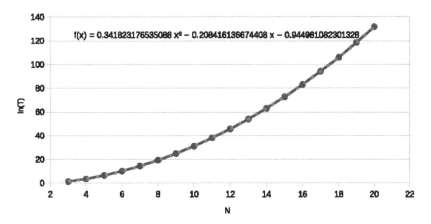

FIGURE 8.6   Logarithm of the number of connected graphs as a function of the number of nodes.

8.7.4 Chromatic Polynomial

```
function chi=chromaticNumber(G)
   % G is a binary adjacency matrix
   % Matlab command:
   P=chromaticPoly(G); % Build the chromatic polynom
   N=length(P);
   % Find the smallest integer chi so that P(chi)>0
   % This is the chromatic number
   for chi=1:N-1 % In fact 1 can not be acceptable for a connected graph
     pnx=evalP(P,N,chi);
     if evalP(P,N,chi)>0 return; end
   end
end
%===========================================================================
function Px=evalP(P,N,x)
Px=0;
u=1;
for n=2:N-1
  u=u*x+P(n);
end
Px=u+P(N);
end
```

8.7.5 Deterministic Algorithms

Some instructions may not work under Octave, especially for graphics.

For guaranteed algorithms (linear programming and sequential search) with necessarily long computing times, in practice .mex C+ files can be generated, compiled under Matlab®. In addition, the codes could be written more concisely, using specific instructions, but I avoided doing so to facilitate a possible translation to other languages.

8.7.5.1 Linear Programming

```
function CC=colorAlgoLP(G) % If called by graphColor
  %{
  Use linear programming under constraints for optimal graph coloring.
  The solution is sure, but it takes a very long time
  when N increases.

  Note: this code is neither elegant nor concise,
  so that it should be easier to understand.
  The only difficult point is how to build
  the matrix A for inequalities constraints.

   G    = adjacency matrix of the graph

  %}
  [N,~]=size(G);
  CC=zeros(1,N);
  if N>30 % Of course, it depends on your computer ...
    fprintf('\n WARNING: It may be looong!'); fprintf('\n')
  end
  K=2; % To start, try this number of colors.
     % You may perform first
     % a structural analysis, in order to find
     % a higher lower bound.
  OK=false;
  while ~OK
   NK=N*K;
   %-------------------------------------------------------------------
   % Build Aeq and beq so that Aeq*x=beq
   % fprintf('\n Build Aeq and beq, for equality constraints');
   beq=ones(N,1);
   Aeq=zeros(N,NK);
   for i=1:N
    j1=K*(i-1);
    for j=j1+1:j1+K
     Aeq(i,j)=1;
    end
   end

   s=sum(G(:));
   A=zeros(s ,NK);

   edg=0;
   for i=1:N-1
    for j=i+1:N
```

```
   if G(i,j)==0 continue; end
   edg=edg+1;
   for k=1:K
    rxi=(i-1)*K + k;
    rxj=(j-1)*K + k;
    A(edg+k-1,rxi)=1;
    A(edg+k-1,rxj)=1;
   end
   edg=edg+K-1;
  end
end

[s,~]=size(A);
b=ones(s,1);

% Now we try to solve
intcon=1:NK; % All xi are integers
lb=zeros(1,NK); % Actually the only acceptable values are 0 and 1
ub=ones(1,NK);

options=optimoptions(@intlinprog,'Display','off');
%options=optimoptions(@intlinprog,'Display','iter');
%  If A is sparse it could be better to use mldivide
% or decomposition.
[x,~,exitflag] = intlinprog([],intcon,A,b,Aeq,beq,lb,ub,options);

if exitflag~=1
 %fprintf('\n No valid coloring with %i colors \n',K)
 K=K+1; % So we try with more colors
else
 % A solution has been found. Build the NxK binary representation
 OK=true;
 C=reshape(x,K,N);
 Ct=C';
 [~,l]=size(Ct);
 % Convert the binary representation into the classical one
 %try
 for i=1:N
  CC(i)=find(Ct(i,1:l)>0);
 end


 %{
 % Check the validity, according to the mathematical proof
 GK=G;
 for k=1:K-1
```

```
 GK=[GK,G]; % size N,N*K
 end
 Sigma=outerSum(C,C);
 Gamma=GK.*Sigma;
 m=max(Gamma(:));
 if m>1
  fprintf('\n Invalid coloring ');
  fprintf('\n There should be no 2 here: \n');
  disp(Gamma);
 end
 %}

 % disp(Aeq)
 % disp(A)

 end
 end
```

8.7.5.2 Greedy 0

```
function C=colorAlgo0(G)
  % Classical Greedy
  fprintf('\n Algo 0');
  [N,~]=size(G);

  sequence=false; % If true plot the successive colorings
          % Just to see what happens.
          % Do NOT use it if N is big!
  % Check
  if sequence && N>10
   fprintf('\n Risk of out of memory')
   error(" ")
  end
  edges=G2edges(G); % Just for plot
  C=zeros(1,N); % Initial colors

  if sequence graphPlot(edges, C); end

  colored=0; % Number of colored nodes

  while colored<N
    for n=1:N
      % List of colors used by its neighbours
      if C(n)>0 continue; end
      L=[];
      for nn=1:N
```

```
      if G(n,nn)==1 && C(nn)>0
        L=[L C(nn)];
      end
    end % for nn=1:N

    if isempty(L)
     k=1;
     else
       % Find the smallest color k that is
       % NOT used by its neighbours
       % and assign it to n
       for k=1:N
        I=find(L==k);
        if isempty(I)
          break;
        end
        end %  for c=1:N
     end % if isempty(L)

    C(n)=k;
     colored=colored+1;
     %fprintf('\n %i colored %i',n,k);
   if sequence graphPlot(edges, C); end

   end %  for n=1:N
  end % while colored<N

end
%-------------------------------------------------
function graphPlot(edges,C)
%{
  Plot a colored graph
   edges = edges of the graph
   C = colors of the nodes. May be empty
Warning: specific to Matlab. Does not work with Octave (2021)
%}
colors=[240 240 240;
 255 0 0;160 160 160;
 255 255 0;0 255 0;
 0 128 255;102 0 204;
 204 153 255; 255 153 153;
 255 255 204;229 255 204;
 153 204 255
 ];
colors=colors/255;
```

```
G=graph(edges(:,1),edges(:,2));
N=length(C);
figure('Color',[1 1 1]); % White background
if ~isempty(C)
  h=plot(G,"-o",'EdgeColor','k');
  h.MarkerSize=30;
  h.NodeFontSize = 20;
  %h.NodeLabel = {}; % Remove the labels of the nodes
  % Coloring
   if N<=length(colors)
    h.NodeColor=colors(C+1,:);
   else
     h.NodeCData=C;
   end

else
  h=plot(G,"-o",'EdgeColor','w');
  h.NodeColor="black";
  h.MarkerSize=32;
   h.NodeLabel = {};
  hold on
  h2=plot(G,"-o",'MarkerSize',30,'EdgeColor','k');
   h2.NodeColor="white";
  %h2.NodeFontSize = 20;
  %h2.NodeFontName="Times";
  %h2.NodeFontAngle="normal";
   h.NodeLabel = {};
  end
  axis off; % Remove the axis
  end
```

8.7.5.3 *Greedy Eccentric*

```
function C=colorAlgoEcc(G)
   % Based on Algo 0 (classical greedy), but more eccentric:
   % it does not always assign the smallest possible color
   fprintf('\n Algo Eccentric');
   [N,~]=size(G);
   C=zeros(1,N); % Initial colors
   C(1)=1;
   colored=1; % Number of colored nodes
   Lused=[1]; % List of colors of colored nodes
   while colored<N
     for n=1:N % For node n
       if C(n)>0 % Already colored
        continue;
```

```
      end

      % List of colors used by its neighbours
      Lneigh=[];
      for nn=1:N
       if G(n,nn)==1 && C(nn)>0
        Lneigh=[Lneigh C(nn)];
       end
      end % for nn=1:N
        % Find the HIGHEST color that is already used
        % but NOT by its neighbours
        % and assign it to n
        Lused=sort(Lused,'descend');
        OK=false;
        for k=1:length(Lused)
         I=find(Lneigh==Lused(k));
         if isempty(I)
          OK=true;
          c=Lused(k);
          break;
         end
        end %  for k=1:length(Lused)

        if ~OK
         c=max(C)+1;
        end

        C(n)=c;
          Lused=[Lused,c];
          colored=colored+1;

     end %  for n=1:N % For node n
    end % while colored<N
   end
```

8.7.5.4 Sober

```
function C=colorAlgoSober(G)
   fprintf('\n Sober (simplification) ');
   %{
     Just for fun. Equivalent to Greedy 0, anyway ..
     We start with the worst possible coloring (N colors)
     and try to simplify.
   %}
   [N,~]=size(G); % G is the incidence matrix (symmetrical)
   C=1:N; % Initial colors
```

```
if sum(G)==N*(N-1) % Fully connected graph
  return;
end
% Find the edges that do not exist
[noEdges,nbNoEdges]=G2edges(1-G - eye(N));
noEdge=0; % Rank of the considered no-edge
while noEdge<nbNoEdges
 noEdge=noEdge+1;
 c1=C(noEdges(noEdge,1)); % Colors of the ends of the no-edge
 c2=C(noEdges(noEdge,2));
 if c1~=c2 % If not the same color
  % try to replace the highest by the smallest
  [cmin, ~]=min([c1,c2]);
  [~, indMax]=max([c1,c2]);

  j=noEdges(noEdge,indMax);
  % Is there a neighbour of j that has the color cmin?
  OK=true;
  for k=1:N
   if G(j,k)== 0; continue; end % No link
   if C(k)==cmin % A link (j-k) and color of k is cmin => no way
    OK=false;
    break;
   end
  end

  if OK
   C(j)=cmin;
  end
 end % if c1~=c2
end % while simplified
C=colorEquiv2(C,N); % To ensure the coloring is permissible
end
```

8.7.5.5 Greedy 1

```
function C=colorAlgo1(G)
  % Greedy. Nodes sorted once by degree
  fprintf('\n Algo 1');
  [N,~]=size(G);
  C=zeros(1,N);
  colored=0;

  % Sort nodes by degree
  for n=1:N
   deg(n)=sum(G(n,1:N));
```

```
end

  [~,indDeg]=sort(deg,'descend');
  while colored<N
    for m=1:N
      n=indDeg(m);

        % List of colors used by its neighbours
        L=[];
        for nn=1:N
          if G(n,nn)==1 && C(nn)>0
            L=[L C(nn)];
          end
        end % for nn=1:N

        if isempty(L)
          k=1;
          else
            % Find the smallest color k that is
            % NOT used by its neighbours
            % and assign it to n
            for k=1:N
              I=find(L==k);
              if isempty(I)
                break;
              end
            end %  for k=1:N
        end % if isempty(L)

        C(n)=k;
        colored=colored+1;
      end %  for n=1:N
    end % while colored<N
```

8.7.5.6 Greedy 2

```
function C=colorAlgo2(G)
    % Greedy. Color the node that has the max number
    % of non colored neighbours
  fprintf('\n Algo 2');
    [N,~]=size(G);
    C=zeros(1,N); % No initial colors
  colored=0; % Number of colored nodes
  while colored<N
    % Find the non colored node n that has the max number
    % of uncolored neighbours
```

```
nMax=0;
for i=1:N % Loop on nodes
  if C(i)>0 continue; end % If already colored, skip
   nbNeigh=0;
   for nn=1:N % Number of non colored neighbours of i
    if G(i,nn)==1 && C(nn)==0
     nbNeigh=nbNeigh+1;
    end

    if nbNeigh>=nMax
     n=i;
     nMax=nbNeigh;
    end
   end
 end
%fprintf('\n Node %i:',n)
    % List of colors used by its neighbours
    L=[];
    for nn=1:N
     if G(n,nn)==1 && C(nn)>0
      L=[L C(nn)];
     end
    end % for nn=1:N

    if isempty(L)
     k=1;
     else
      % Find the smallest color k that is
      % NOT used by its neighbours
      % and assign it to n
      for k=1:N
       I=find(L==k);
       if isempty(I)
        break;
       end
      end %  for c=1:N
    end % if isempty(L)
%fprintf(' assign color %i',k);
    C(n)=k;
    colored=colored+1;
 end
  end
```

8.7.5.7 Greedy 3

```
function C=colorAlgo3(G)
    % Greedy less sensitive to the numbering of the nodes
    fprintf('\n Algo 3');
    [N,~]=size(G);
    sequence=false; % If true plot the successive colorings
              % Just to see what happens.
              % Do NOT use it if N is big!
    % Check
    if sequence && N>10
      fprintf('\n Risk of out of memory')
      error(" ")
    end
    edges=G2edges(G); % Just for plot
     C=zeros(1,N); % Initial colors
      C(1)=1;
     colored=1; % Number of colored nodes
     coloredNodes=1;
     degr=sum(G,1);
     [~,permut]=sort(degr);
    if sequence graphPlot(edges, C); end
     while colored<N
       % Consider the set of colored nodes
       % Find a non colored node n that is connected to this set

      % for n=1:N % Loop on nodes
      for n=1:N
        if C(n)>0 continue; end % Already colored => ignore

         % Let Nk be the subset of colored nodes connected to n
         Nk=[];
         for k=1:colored
          c=coloredNodes(k);
         if G(c,n)==1
           Nk=[Nk c];
         end
         end

        if ~isempty(Nk)
        Ck=C(Nk);

          % Find the smallest color colorK that is not used in Nk
          for colorK=1:N
           used=find(Ck==colorK);
           if isempty(used)
```

```
            % Assign colorK to n.
            C(n)=colorK;
            break
            end
          end  % for colorK=1:N

        coloredNodes=[coloredNodes n];
        colored=colored+1;
        if sequence graphPlot(edges, C); end
        end %  if ~isempty(Nk)
      end % for n=1:N % Loop on nodes

    end % while colored<N
    end
```

8.7.5.8 Greedy 4

```
function C=colorAlgo4(G)
    fprintf('\n Algo 4');
    [N,~]=size(G);
    sequence=true; % If true plot the successive colorings
            % Just to see what happens.
            % Do NOT use it if N is big!
    % Check
    if sequence && N>10
      fprintf('\n Risk of out of memory')
      error(" ")
    end
    edges=G2edges(G); % Just for plot
    % Depth indices
    for i=1:N-1
      for j=i+1:N
        depth(i,j)=pathG(G,i,j);
      end
    end
    % Initialisation
    C=zeros(1,N) ; % Initial colors
    k=1;
    colored=0;

    if sequence graphPlot(edges, C); end

    while colored<N
    % Find the max depth(i,j) with at least i or j not colored
    minDepth=Inf;
    for i=1:N-1
```

```
  for j=i+1:N
   if C(i)*C(j)>0 continue;end % Both nodes are colored
    if depth(i,j)<minDepth
     minDepth=depth(i,j);
    if C(i)==0
     n=i; % Node to color
    else % C(j)=0
     n=j;  % Node to color
    end
   end
  end
 end

 % List of colors used by its neighbours
    L=[];
    for nn=1:N
     if G(n,nn)==1 && C(nn)>0
      L=[L C(nn)];
     end
    end % for nn=1:N

    if isempty(L)
     k=1;
    else
      % Find the smallest color k that is
      % NOT used by its neighbours
      % and assign it to n
      for k=1:N
       I=find(L==k);
        if isempty(I)
         break;
        end
      end %  for k=1:N
    end % if isempty(L)

    C(n)=k;
    colored=colored+1;
    if sequence graphPlot(edges, C); end

 end % while colored<N
 end
%========================================================
function l=pathG(G,i,j)
% Length of the shortest path between i and j
% The graph G is connected and undirected
[N,~]=size(G);
```

```
if G(i,j)==1
 l=1;
 return
end
% Neighbours
% j is not amongst them
l=2;
 neigh1=[i];
 used=zeros(1,N);
 used(i)=1;
 n1=i;

 isj=false;
 while ~isj
  l1=length(neigh1);
  neigh2=[];

  for n2=1:l1
   for  k=1:N
    if G(neigh1(n2),k)==1
     if k==j
      isj=true;
      break;
     else
      if used(k)==0
       neigh2=[neigh2 k];
       used(k)=1;
      end
     end
    end
   end % for  k=1:N

   if isj
    l=l-1;
    return
   end

  end  % for n2=1:l1
  l=l+1;
  neigh1=neigh2;
 end % while ~isj
end
```

8.7.5.9 Johnson

```
function C = colorAlgoJohnson(G)
   % Thanks to Stephan (2022)
   % Warning: this Matlab code may not work for Octave
   fprintf('\n Johnson')
    edges=G2edges(G);
    s=edges(:,1);
    t=edges(:,2);
   Graph = graph(s,t);
   N=numnodes(Graph);
   % Initialisation
     Ccurrent = 0;
     C = zeros(1,N )';
     C(:) = Inf; % To start, set the colors to infinity
     node = cell(N,1);
     for n = 1:N
       node{n} = num2str(n);
     end
     Graph.Nodes.Name = node;
     GraphNew = Graph;
   %--------------------------------------------------- Algorithm
     while sum(isinf(C)) > 0 % While there are non colored nodes
       Ccurrent = Ccurrent + 1;
       GraphNew = Graph;
       % Remove the colored nodes. It creates a subgraph
       GraphNew = rmnode(GraphNew, find(~isinf(C)));
       while numnodes(GraphNew) > 0 % If there still are
                      % non colored nodes
         % Find the node of the subgraph  of uncolored nodes
         % that has the smallest degree
         [~, nodeIndex] = ...
         min(degree(GraphNew, GraphNew.Nodes.Name));
         % Assign the current color
         C(str2double(cell2mat(GraphNew.Nodes.Name(nodeIndex))))...
           = Ccurrent;
         % Find the neighbours of these nodes
         neigh = neighbors(GraphNew,...
           (cell2mat(GraphNew.Nodes.Name(nodeIndex))));

         % Remove this node
         GraphNew = ...
         rmnode(GraphNew, GraphNew.Nodes.Name(nodeIndex));
         % Remove its neighbours
         GraphNew = rmnode(GraphNew, neigh);
       end
```

```
   end
  end
  %========================================================
  function edges=G2edges(G)
  [N,~]=size(G); % Number of nodes (connected symmetric graph)
  nE=N*(N-1)/2;
  edges=zeros(nE,2);
  nbEdges=0;
  for i=1:N-1
   for j=i+1:N
    if G(i,j)>0
     nbEdges=nbEdges+1;
     edges(nbEdges,1)=i;
     edges(nbEdges,2)=j;
    end
   end
  end
  edges=edges(1:nbEdges,:);
  end
```

8.7.5.10 RLF

```
%{
   See  MRLF
   https://hal.archives-ouvertes.fr/hal-00451266
   %}
   function C=colorAlgoRLF(G)
   fprintf('\n Algo RLF');
   [N,~]=size(G);
   degr=sum(G,2);
   [~,Ind]=sort(degr,'descend');
   C=zeros(1,N);  % Initial colors = no color at all
   colored=0; % Number of colored nodes
   k=1;
   % Assign color k
   while colored<N
    add=false;
    for n=1:N
     m=Ind(n);
     if C(m)>0 continue; end % Already colored

     % Uncolored node, check the neighbours
     possible=true;
     for nn=1:N
      if nn==m continue; end
      if  G(m,nn)==1 && C(nn)==k % A neighbour is colored with k
```

```
    possible=false;
    break;
   end
  end % for nn=1:N

  if possible % No neighbour is colored with k
   C(m)=k;
   colored=colored+1;
   add=true;
  end
 end % for n=1:N

 if ~add
  k=k+1;
 end
end
end
```

8.7.5.11 Backtracking

```
function C=colorAlgoBack(G)
  %{
  Recursive coding.
  %}
  fprintf('\n Backtracking')
  [N,~]=size(G);
  K=2;
  OK=false;
  iterK=0; % Just for information
  while ~OK
   iterK=iterK+1; fprintf('\nK %i,  %i',K,iterK)
  C=zeros(1,N);
  C(1)=1;
  j=1;
  C=Color(C, j,K,N,G);
  Ind=find(C==0);
  if isempty(Ind)
   OK=true;
  else
   K=K+1;
   iterK=0;
  end
  end
  end
  %======================================================
  function [C]=Color(C, j,K,N,G)
```

```
if j==N+1
 return;
end
for i=1:K
 C(j)=i;
 if valid(C,G,j)
 [C]= Color(C,j+1,K,N,G);
 break;
 end
end
end
```

8.7.5.12 Sequential Search

```
function C=colorAlgoSeqImprov2 (edges,K0)
 %{
 Garanteed version, but time consuming
 K0 = initial number of colors to try.
 Then the algorithm may increase or decrease it.
 %}
 saveSkip=false;
 if saveSkip skipFile=fopen('skip','w'); end
 saveColor=false;
 if saveColor saveC=fopen('saveC','w'); end
 verbose=false;
 N=max(edges(:));
 [nbEdges,~]=size(edges);
 valid=false;
 if  K0==0
  u=nbEdges/(N*(N-1)/2);
  K=floor(max(2,N*u^(N/2)));% Try first
                % this number of colors
 else
  K=K0;
 end
 decreaseK=true;
 admissible=1;
 %skip=[]; % For information. Number of inadmissible
      % after each admissible
 nskip=0;
 END=false;
 C=[zeros(1,N-K+1), 1:K-1]; % Cmin. Admissible
 Cmax=[0,(K-1)*ones(1,N-1)]; %
 while 1==1 % Seems infinite, but one always finds
      % a solution and return
  if K==N  % For the complete graph
```

```
 Cbest=1:N;
 return
end

% Check validity
valid=checkValid2(edges,nbEdges,C);

if valid
 if decreaseK
  if K>2
   K=K-1;
   C=[zeros(1,N-K+1), 1:K-1]; % Cmin. Admissible
   Cmax=[0,(K-1)*ones(1,N-1)]; %
  else % K=2, and valid => END
   END=true;
  end
 else % Increasing K and valid => END
  END=true;
 end
else % If not valid
 if sum(C==Cmax)==N % All coloring with K
          % have been checked
  K=K+1; % I increase K
  decreaseK=false;
  C=[zeros(1,N-K+1), 1:K-1]; % New Cmin. Admissible
  Cmax=[0,(K-1)*ones(1,N-1)]; %

 else % Try the next coloring with K
  [C,skipped]=nextSkip2(C,N,K);
  if skipped % Either K=2 or skip possible
   admiss=true;
  else % Check admissibility
   admiss=checkAdmiss(C,K);
   if admiss
    admissible=admissible+1;
   end
  end

  if admiss
   % Save C
   if saveColor
    for n=1:N
     fprintf(saveC,'%i ',C(n))
    end
    fprintf(saveC,'\n')
   end
```

```
    %skip=[skip; nskip]; nskip=0; % For information
   else
    nskip=nskip+1; % Interval between
            % successive admissibles
   end
  end
 end

 if END
  if saveSkip
   for s=1:length(skip)
    fprintf(skipFile,'%i \n',skip(s));
   end
   fclose(skipFile);
  end
 C=C+1;
  return %***** END
 end

end
end
%==========================================================
function admiss=checkAdmiss(C,K)
admiss=true;
for k1=1:K
 isk=C==k1-1;
 if isempty(isk) % One colour in (0,..., K-1) is missing
  admiss=false;
  return
 end
end
%==========================================================
function valid=checkValid2(edges,nbEdges,C)
valid=false;
for i=1:nbEdges
  if C(edges(i,1))==C(edges(i,2))
   return
  end
end
valid=true;
end
%==========================================================
function [C,skipped]=nextSkip2(C0,N,K)
% Note there is a repetition of nextC
C=zeros(1,N);
skipped=false;
```

```
if K==2
% C=nextC(C0,N,K);
 n=N;
 stop=false;
 C=C0;
 while ~stop
  C(n)=C(n)+1;
  if C(n)==K
   C(n)=0;
   n=n-1;
   if n==0
     stop=true;
   end
  else
    stop=true;
  end
 end
 %----------
 skipped=true; % Actually skip=0, but the coloring
       % is admissible for sure
else
 C1=[zeros(1,N-K+1), K-1:-1:1];
 if sum(C0==C1)<N
 % C=nextC(C0,N,K);
 n=N;
 stop=false;
 C=C0;
 while ~stop
  C(n)=C(n)+1;
  if C(n)==K
   C(n)=0;
   n=n-1;
   if n==0
     stop=true;
   end
  else
    stop=true;
  end
 end

 %----------------
 skipped=false;
else % We can directly skip to C2
 C=[ zeros(1,N-K), 1,0,2:K-1];
 skipped=true;
end
```

```
   end
end


8.7.6  Bi-objective

function C=colorAlgoBiobj(G0,edges,nbEdges)
   global G % To avoid nested functions
   G=G0;
   [N,~]=size(G); % Number of nodes
   lb=ones(1,N);
   ub=N*lb;
   options = ...
   optimoptions('gamultiobj','ParetoFraction',0.7,...
   'PlotFcn',@gaplotpareto);
   [solution,f12] = ...
   gamultiobj(@biobj,N,[],[],[],[],lb,ub,options);
   solution=round(solution); % List of all solutions
   % Find a valid solution, if any
   [nbSol,~]=size(solution);
   Cvalid=[];
   nbValSol=0;
   for s=1:nbSol
      Cs=colorEquiv2(solution(s,:),N); % Transform the coloring
                     % into a minimal equivalent one
       valid=...
      checkValid2(edges,nbEdges,Cs); % True iif each arc is valid
      if valid
          Cvalid=[Cvalid;Cs];
          nbValSol=nbValSol+1;
          f1(nbValSol)=...
          f12(s,1); % Number of colors of this coloring
      end
   end
   if isempty(Cvalid)
      fprintf("\n Warning: invalid colouring!\n")
      C=Cs(1,:);
       return
   end
   % Select the valid solution with
   % the minimum number of colors
   [~,Ind]=sort(f1);
   C=Cvalid(Ind(1),:);
   end
   %--------------------
   function f=biobj(x,G)
   global G
```

```
C=round(x);
N=numel(C);
% Number of colors, to minimise
f(1)=numel(unique(C));
% Number of invalid arcs, to minimise
nbInval=0;
for i=1:N-1
    for j=i+1:N
        if G(i,j)>0 && C(i)==C(j)
            nbInval=nbInval+1;
        end
    end
end
f(2)=nbInval;
end
```

8.7.7 Quantum Algorithm

The code is in Qiskit (Qiskit 2022) simulation language.

```
'''
  NK hybrid method for the K-coloring of a N-graph
  Maurice.Clerc@WriteMe.com
  2023-07
  ---
  Adjacency matrix for the graph G
  Binary coloring matrix:
  1 row for each node
  1 column for each color
  One and only one 1 in each row, other values 0
  It means we need NK qubits to describe a coloring matrix
  Steps
  1) W state to define a superposition of states that respects the above
     constraints
  2) Use one ancillary qubit for each edge, set it to |1> if valid
     (different colours on the ends)
  3) Use one ancillary qubit set to |1> if all edges are valid
  4) Define the marker/filter/oracle by the condition "this qubit is
     in state |1>"
  5) Use it for a Grover's search. Note that we just need to find _one_
     valid coloring, if possible.
     so you may try to force the number of iterations to a value smaller
     than the optimal one.
  ---
  On my laptop I can consider only very small graphs (at most 10 qubits nodes).
  For bigger graphs the code runs on the IBM quantum simulator
  (https://quantum-computing.ibm.com/)
```

It means you need to be registered (for free) and to have an API code.
Used here under Anaconda / Jupyter /Python / Qiskit
Warnings:
1) The algorithm searches a solution for a given number of colors (K).
For a complete resolution it should loop on different K values.
The worst approach is to check for K=2, then 3, ... N-1,
but clever strategies do exist (dichotomy, for instance)
2) This Qiskit code is not the most efficient one, but should be easy
  to understand
'''

```python
import math
import numpy as np
from qiskit import *
import qiskit.tools.jupyter
from qiskit.quantum_info import Statevector
from qiskit.visualization import plot_histogram
%matplotlib inline
nQbitsMax=10 # maximum number of qubits my laptop can safely manipulate
                # If more, I use the IBM cloud
# Describe the graph as a binary sequence
# (upper right half-triangle of the adjacency matrix)
# read row by row from left to right
# Graphs for tests
G2_1_2=[1] # 2 nodes,one edge, needs 2 colors
G3_2_2=[1,0,1] # Linear 3 nodes,2 edges, needs 2 colors
G3_3_3=[1,1,1 ] # 3 nodes, 3 edges,needs 3 colors
G4_3_2=[1,0,0,1,0,1] # Linear 4 nodes, 3 edges, needs 2 colors
G4_4_3=[1, 0, 0, 1, 1, 1] #  4 nodes, 4 edges, needs 3 colors
G4_5_3=[0, 1, 1, 1, 1, 1] #  4 nodes, 5 edges, needs 3 colors
G5_5_3=[1,0,0,0,1,0,0,1,1,1]  #5 nodes,5 edges, needs 3 colors
G5_8_4=[1,1,0,1,1,0,1,1,1,1]  #5 nodes, 8 edges, needs 4 colors
# Complete N nodes graph
N=4 # For N>=5 => error "'I' format requires 0 <= number <= 4294967295"
Gc=[1] * round(N*(N-1)/2)
# Select the graph and the number of colors to try
G=G4_4_3 # Any above graph or Gc
nColors=3 # Should be at most nNodes. Set it to N if G=Gc
toDraw=False
toList=False # to list all the solutions, not only the first one
                # Warning: this list can be very long.
                # For the complete N-graph it is N^N ...
#------------------
l=len(G)
nNodes=round(1/2 +math.sqrt(2*l+1/2))
nEdges=sum(G)
#------------------------------------- Draw the graph
```

```python
if toDraw: # Create a graph from the adjacency matrix and draw
import networkx as nx
import matplotlib.pyplot as plt
adjMatrix=np.zeros((nNodes, nNodes))
ij=0
for i in range(nNodes):
    for j in range(i+1,nNodes):
    adjMatrix[i][j]=G[ij]
     ij=ij+1
Gdraw = nx.from_numpy_matrix(adjMatrix)
pos = nx.circular_layout(Gdraw)
nx.draw_networkx(Gdraw, pos=pos, with_labels=True, node_color='lightgrey', )
plt.show()
#------------------------------------ Build the circuit
print([nNodes,' nodes'])
nnc=nNodes*nColors # For W-state coloring matrix
nqbits=nnc+nEdges+1 # Total number of qubits.
print([nqbits,"nqbits"])
print([nEdges, 'edges'])
# Create a Quantum Circuit
q = QuantumRegister(nqbits)
circ = QuantumCircuit(q)
#--------- Generate a superposition of all coloring matrices
# Initialisation
# W-state. Describe the coloring matrix (one 1 and only one by row)
start=0
for n in range(nNodes):
circ.x(start+nColors - 1)
for k in range(nColors - 1, 0, -1):
    theta = np.arccos(np.sqrt(1 / (k + 1)))
    circ.ry(-theta, start+k - 1)
     circ.cz(start+k, start+k - 1)
     circ.ry(theta, start+k - 1)
for k in range(1, nColors):

    circ.cx(start+nColors - k - 1, start+nColors - k)

start=start+nColors
#print(dict(circ.count_ops())) # Number of gates of each type
if toDraw:

    circ.draw(output='latex' )

# Just to check the result.
# Each binary sequence has to be read from right to left
# For the moment only the nNodes subsequences of nColors are set
```

```
#sv = Statevector.from_instruction(circ)
#print(len(sv))
#sv.draw('latex')
#------------ Check conflicts (same color at the ends of an edge)
n=-1
target=nnc
for n1 in range(nNodes):

    qnode1=n1*nColors # rank of the first qubit for the node n1
    for n2 in range(n1+1,nNodes):

        qnode2=n2*nColors # rank of the first qubit for the node n2
        n=n+1
        if G[n]>0: # There is an edge
        for k in range(nColors): # For each color k ...

            nk1=qnode1+k # |1> if node n1 has the color k
            nk2=qnode2+k # |1> if node n2 has the color k
            circ.ccx (nk1,nk2,target) # |1> if invalid edge
                            # (same color on the ends)
        target=target+1

begin=nnc
end=nnc+nEdges
circ.x(list(range(begin,end))) # Set to |1> the edges that are valid
print('end of check conflicts')
# The coloring is valid iff all edges are valid (state |1>)
# In that case, set the last (ancillary) qubit to |1>
cb=list(range(start,end))
circ.mcx(cb,end)
print(dict(circ.count_ops())) # Number of gates of each type
print("Circuit depth: ",circ.depth())
if toDraw:
    circ.draw(output='latex' )
# Now, for the valid colorings, the two leftmost bits are 1
#sv = Statevector.from_instruction(circ)
#sv.draw('latex')
#----------------------------------- Try to find a solution
# Note that classical Grover's method doesn't work if half the possible strings
# are solutions
# so in such a case (here for nNodes=2) we have to "cheat"
oracle=QuantumCircuit(nqbits)
good_state=[end] # The leftmost qubit must be in state |1>
# Note: in that case we know that the second leftmost is also in state |1>
# Actually we also know that if there is a solution
# for at least one the first qubit is in state |1>
```

```python
# Marker/filter/oracle
if nNodes==2:
    control=0
else:
    control=end-1
oracle.h(end)
oracle.cx(control,end)
oracle.h(end)
# Amplify the tagged state,
from qiskit.algorithms import AmplificationProblem
problem = AmplificationProblem(oracle,state_preparation=circ, \
        is_good_state=good_state)
# You can plot the Grover circuit
groverCirc=problem.grover_operator.decompose();
print(dict(groverCirc.count_ops())) # Number of gates of each type
#sv = Statevector.from_instruction(groverCirc)
#print(len(sv))
#print(sv)
#sv.draw(output='latex')
if toDraw:
    #groverCirc.draw(output='mpl' )
    groverCirc.draw(output='latex' )
    #print(problem.grover_operator.decompose())
# Just to evaluate the complexity
d=circ.depth()
print("Circuit width: ",nqbits)
print("Circuit depth: ",d)
groverDepth=problem.grover_operator.decompose().depth()
print("Grover's circuit depth:",groverDepth)
from qiskit.algorithms import Grover
if nqbits<nQbitsMax: # Can run locally on a small laptop
    from qiskit.primitives import
    # Force the number of iterations
    #grover = Grover(iterations=100,sampler=Sampler())
    grover = Grover(sampler=Sampler()) # Automatically estimated
else: # IBM quantum cloud.
        # The needed code has been saved once on the local computer
    from qiskit_ibm_runtime import Sampler
    from qiskit_ibm_runtime import QiskitRuntimeService
    service = QiskitRuntimeService(channel="ibm_quantum")
    backend = service.backend("ibmq_qasm_simulator")
    grover = Grover(sampler=Sampler(backend=backend))
result = grover.amplify(problem)
print("Assignment",result.assignment)
print("Iterations",result.iterations)
print("Max probability",result.max_probability)
```

```python
print("Validity (not always trustable)",result.oracle_evaluation)
print("Top measurement",result.top_measurement)
nbIter=len(result.iterations)
# Because just one marked state:
nbIterTheor=math.floor((math.pi / 4) * math.sqrt(2**nqbits))
print('Number of iterations', nbIter)
if nnc<10: # The histogram is readable only for very small graphs

    display(plot_histogram(result.circuit_results[0]))
#---------------- Post processing
binary=result.top_measurement
failure= binary[0]=='0'
if failure:
    print("Sorry, I didn't find any solution")
else:
    print('Here is the first valid coloring:')
reverse_str = binary[::-1] # Should be read from right to left
print("Binary coloring solution")
Color=[]
for i in range(nNodes):
start=i*nColors
end=start+nColors
row = reverse_str[start : end]
print(row)
for j in range(nColors):
    if int(row[j])>0:
        Color.append(j+1)
print("Color list (by node)")
print(Color)
print(" ")
if toDraw: # Draw the colored graph
    nx.draw_networkx(Gdraw, pos=pos, with_labels=True, node_color=Color)
    plt.show()
# Valid colorings list (only for simulation)
r=result.circuit_results # This a list ... with just one element
r0=r[0] # This is a dictionary
#print('Note:there are at least',math.factorial(nNodes) , 'solutions')
#print(len(r[0]),"Found")
if nColors>nNodes:
    print('(because more colors than nodes)')
x=r0.keys()
binaries=list(x)
nbValid=0
for i in range(len(binaries)):
    if binaries[i][0]=='1':
        nbValid=nbValid+1
```

```
    if toList:
        print(binaries[i])
print('Valid colorings:',nbValid)
```

8.7.8 Diplomate 0

An Octave/Matlab code deliberately not optimized, in order to highlight the steps of the method.

```
function C=colorAlgoDiplomat0(G,frust,nbColors)
% frust= NxN values (one for each arc)
%     Not the most compact way, but easier to use
fprintf('\n =====================================Diplomat 0');
[N,~]=size(G);
frustTot=0;
frustMax=sum(frust(:));
edges=G2edges(G); % Just for plot
C=[1,zeros(1,N-1)]; % Initial colors
coloredNb=1; % Number of colored nodes
while coloredNb<N
 frstMin=Inf;
 for i=1:N % For each uncolored node
  if C(i)>0 continue; end
  % List of colored neighbours
  coloredNeigh=[];

  for j=1:N
   if G(i,j)==0 continue; end % No link
   if C(j)==0 continue; end % A link, but the node is uncoloured
   coloredNeigh=[coloredNeigh,j];
  end

  % Try colors
  for k=1:nbColors
   % If assigned to the node, compute the generated frustration
   frst=0;
   for neigh=1:length(coloredNeigh)
    j=coloredNeigh(neigh);
    if C(j)==k
     frst=frst+frust(i,j)+frust(j,i);
    end
   end

   % Keep the best pair (node, color)
   if frst<frstMin
    frstMin=frst;
    iBest=i;
```

```
      kBest=k;
     end
    end
   end % for n=1:N
   fprintf('\n Node %i colored by %i => frustration %f',iBest,kBest,frstMin)
   frustTot=frustTot+frstMin;

   C(iBest)=kBest;
   coloredNb=coloredNb+1;
  end % while coloredNb<N
  fprintf("\n Total frustration %f/%f = %f",frustTot, frustMax,...
   frustTot/frustMax);
  end
```

8.8 QUANTUM RACE

This little board game is vaguely inspired by quantum circuits, hence its name. We can imagine that each token represents a qubit and the color its state: white for $|0>$ and black for $|1>$. And these tokens go through doors (gates) that transform them.

8.8.1 Material

Figure 8.7 shows a "luxury" wooden version, in which the nine-hole bars are used to count the number of games won, but the whole game can be made in cardboard and the scores simply noted on paper.

- A playing area, on which are drawn the tracks or time lines (horizontal, one per player) and the instant lines (vertical, odd number, for example, from 0 to 6 for single games).

- For each player, a black and a white token or, better yet, a black-white biface. You can also use more tokens, to leave them in place and better visualize the game when finished (see examples below).

- A set of wafers or cards for the "quantum gates," each in several copies:

  1. X gate. It switches the token that crosses it (white gives black and vice versa).

  2. H gate. After crossing, the color of the token is randomly drawn.

  3. CX gate (C for control). At the moment, if the player's token is black, it gives him the right to replace the gate of an opponent at the same time by an X gate, drawn separately.

  4. I (identity gate). The token passing through remains the same.

Any device to make a random binary choice (heads, black-and-white dice, etc.).

TABLE 8.1 Quantum Race, Example 1

	0	1		2		3		4		5		6	
Sam	○	X	•	I	•	H	○	H	○	CX	○	X	•
Sara	○	X	•	H	○	I	○	H	•	X	○	CX	○

8.8.2 Rules and Conduct of the Game

Here, we assume that there are 7 time lines, from 0 to 6.

- At the beginning, each player places their white token on his track, at 0. The goal of the game is to arrive with a black token at 6.

- 6 gate cards, face down, are randomly distributed to each player.

- A special package of 6 X gates is placed separately.

- Each turn, where players each play one shot, is a one-time step progression.

- Each player chooses a gate in their game and poses it face down in front of him. When all the players do it, these cards can be turned, each player places their own on their track, on the current time step.

- If a player has presented a CX gate and their token is currently black (implied to be "active"), they replace the gate of another player with an X taken from the special package, unless that gate is also a CX.

- Then, each player moves their token through his current gate, applying the corresponding transformation:
  - For an I gate, the color doesn't change.
  - For a CX gate, the color doesn't change.
  - For an X gate, the color is switched.
  - For an H gate, the color is drawn at random.

8.8.3 Example 1, with Two Players

Sam receives the following cards: X I X CX H H . He plays on the first track.

Sara receives the I CX H H X X cards. She plays on the second track.

Table 8.1 presents a complete game, where we left in place the successive tokens, to better see the evolution.

- Time 1. Each player places an X gate. As they pass through, the tokens turn black.

- Time 2. Sam places an I gate. His token will remain black. Sara places an H gate, randomly draws the color and finds white.

- Time 3. Sam places an H gate, randomly draws and finds white. Sara places an I gate. Her token remains white.

TABLE 8.2   Quantum Race, Example 2

	0	1		2		3		4		5		6	
Sam	○	X	•	I, X	•○	H	•	H	•	CX	•	X	○
Sara	○	X	•	CX	•	H	○	H	•	I,X	•○	X	•

FIGURE 8.7   Quantum race, "luxury" version.

- Time 4. Sam places an H gate, randomly draws and finds white. Sara places an H gate, randomly draws and finds black.

- Time 5. Sam places a CX. As his token is white, it will actually be inoperative on Sara. His own token does not change and remains white. Sara places an X. So her token turns white.

- Time 6. Sam places his last gate, an X. His token turns black. Sara places her last gate, a CX that is useless. Her token remains white. Sam wins.

8.8.4 Example 2, with Activated CX Gates

Table 8.2 presents another game, Sam and Sara having the same cards as in the previous example. Here, Sara wins.

- Time 1. Sam and Sara both place an X door. Their tokens will turn black as they pass through it.

- Time 2. Sam placed an I, but Sara placed a CX. As her token is black, she picks an X from the special package and replaces Sam's I gate. His token will turn white as it passes through this X gate.

- Time 3. Sam places an H gate, randomly draws, and finds black. Sara places an H gate, randomly draws, and finds white.

- Time 4. Sam places an H gate, randomly draws, and finds black. Sara places an H door, randomly draws, and finds black.

- Time 5. Sara has placed an I, Sam a CX. As his token is black, he uses a special X card to replace Sara's I. Sara's token will turn white, but it will be actually favorable.

- Time 6. Sam has to put down his last gate, which is an X. His token goes through the gate and becomes white. Sara also puts down her last gate, an X. Her token goes through the gate and turns black. Sara wins.

References

[1] Ahmed, Shamim (2022). "Applications of Graph Coloring in Modern Computer Science." In: *International Journal of Computer and Information Technology* 3.2, pp. 1–7. URL: https://www.academia.edu/2479839/Applications_of_Graph_Coloring_in_Modern_Computer_Science (visited on 10/01/2022).

[2] Ahmadi, Saman et al. (July 2021). "Bi-Objective Search with Bi-directional A*." In: *Proceedings of the International Symposium on Combinatorial Search* 12.1. Number: 1, pp. 142–144. URL: https://ojs.aaai.org/index.php/SOCS/article/view/18563 (visited on 10/01/2022).

[3] Aspect, Alain, Jean Dalibard, and Gérard Roger (Dec. 1982). "Experimental Test of Bell's Inequalities Using Time-Varying Analyzers." In: *Physical Review Letters* 49.25. Publisher: American Physical Society, pp. 1804–1807. DOI: 10.1103/PhysRevLett. 49.1804. URL: https://link.aps.org/doi/10.1103/PhysRevLett.49.1804 (visited on 10/13/2022).

[4] Bilu, Yonatan (2006). "Three Extensions of Hoffman's Bound on the Graph Chromatic Number." In: *Journal of Combinatorial Theory* Series B 96.4, pp. 608–613.

[5] Brélaz, Daniel (Apr. 1979). "New Methods to Color the Vertices of a Graph." In: *Communications of the ACM* 22.4, pp. 251–256. ISSN: 0001-0782. DOI: 10.1145/359094. 359101. URL: https://doi.org/10.1145/359094.359101 (visited on 01/10/2022).

[6] Clerc, Maurice (2007). *When Nearer is Better*. Tech. rep. Open archive. URL: https://hal.archives-ouvertes.fr/hal-00137320.

[7] Clerc, Maurice (Apr. 2015). *L'aléatoire contrôlé en optimisation*. ISTE (International Scientific and Technical Encyclopedia). URL: http://iste-editions.fr/products/l-aleatoire-controle-en-optimisation (visited on 04/21/2015).

[8] DIMACS Graphs and Best Algorithms (2022). URL: http://cedric.cnam.fr/porumbed/graphs/ (visited on 01/12/2022).

[9] Edwards, C. S., and C. H. Elphick (Jan. 1983). "Lower Bounds for the Clique and the Chromatic Numbers of a Graph." In: *Discrete Applied Mathematics* 5.1, pp. 51–64. DOI: 10.1016/0166-218X(83)90015-X. URL: https://www.sciencedirect.com/science/article/pii 0166218X8390015X (visited on 11/07/2021).

[10] Fabrikant, Alex, and Tad Hogg (2002). *Graph Coloring with Quantum Heuristics*. Edmonton, AB.

[11] Gondran, Alexandre, and Laurent Moalic (Dec. 2018). "Optimality Clue for Graph Coloring Problem." URL: https://hal.science/hal-01952598 (visited on 01/25/2023).

DOI: 10.1201/9781003477785-9

[12] Gualandi, Stefano, and Federico Malucelli (Feb. 2012). "Exact Solution of Graph Coloring Problems via Constraint Programming and Column Generation." In: *INFORMS Journal on Computing* 24.1. Publisher: INFORMS, pp. 81–100. DOI: 10.1287/ijoc.1100.0436. URL: https://pubsonline.informs.org/doi/abs/10.1287/ijoc.1100.0436 (visited on 01/12/2022).

[13] Guignard, Adrien (Dec. 2011). "Jeux de coloration de graphes." Thèse de doctorat. Bordeaux 1. URL: http://www.theses.fr/2011BOR14391 (visited on 12/17/2021).

[14] Johnson, David S. (1974). "Worst Case Behaviour of Graph Coloring Algorithms." In: Proceedings 5th Southeastern Conference on Combinatorics, Graph Theory, and Computing (rev. ed.), Winnipeg, MB: Utilitas Math, pp. 513–527.

[15] Leighton, F. T. (1979). *A Graph Coloring Algorithm for Large Scheduling Problems.* National Bureau of Standards. URL: http://archive.org/details/jresv84n6p489 (visited on 01/10/2022).

[16] Lewis, R. M. R. (2016). *A Guide to Graph Colouring.* Springer International Publishing. DOI: 10.1007/978-3-319-25730-3. URL: http://link.springer.com/10.1007/978-3-319-25730-3 (visited on 10/14/2022).

[17] Lewis, R. M. R. (2021). "Algorithm Case Studies." In: *Guide to Graph Colouring: Algorithms and Applications.* Ed. by R. M. R. Lewis. Texts in Computer Science. Cham: Springer International Publishing, pp. 113–154. DOI: 10. 1007/978-3-030-81054-2_5. URL: https://doi.org/10.1007/978-3-030-81054-2_5 (visited on 03/11/2022).

[18] Mazauric, Dorian (2016). *Graphes et Algorithmes – Jeux grandeur nature.* fr. Pages: 1. Inria. URL: https://hal.inria.fr/hal-01366804 (visited on 10/19/2022).

[19] Moreau, Paul-Antoine et al. (July 2019). "Imaging Bell-Type Nonlocal Behavior." In: *Science Advances* 5.7. Publisher: American Association for the Advancement of Science, eaaw2563. DOI: 10.1126/sciadv.aaw2563. URL: https://www.science.org/doi/10.1126/sciadv.aaw2563 (visited on 10/13/2022).

[20] PSC (n.d.). "Particle Swarm Central." URL: http://particleswarm.info.

[21] Qiskit (2022). URL: https://qiskit.org/ (visited on 03/13/2022).

[22] Rauch, Dominik et al. (Aug. 2018). "Cosmic Bell Test Using Random Measurement Settings from High-Redshift Quasars." In: *Physical Review Letters* 121.8. Publisher: American Physical Society, p. 080403. DOI: 10.1103/PhysRevLett.121.080403. URL: https://link.aps.org/doi/10.1103/PhysRevLett.121.080403 (visited on 10/13/2022).

[23] Shimizu, Kazuya, and Ryuhei Mori (June 2022). "Exponential-Time Quantum Algorithms for Graph Coloring Problems." In: *Algorithmica.* DOI: 10.1007/s00453-022-00976-2. URL: https://doi.org/10.1007/s00453-022-00976-2 (visited on 10/01/2022).

[24] Thadani, Satish, Seema Bagora, and Anand Sharma (Jan. 2022). "Applications of Graph Coloring in Various Fields." In: *Materials Today: Proceedings.* 3rd International Conference on "Advancement in Nanoelectronics and Communication Technologies" (ICANCT-2022), 66, pp. 3498–3501. DOI: 10.1016/j.matpr.2022.06.392. URL: https://www.sciencedirect.com/science/article/pii/S2214785322044108 (visited on 10/01/2022).

[25] Ulloa, Carlos Hernández et al. (June 2020). "A Simple and Fast Bi-Objective Search Algorithm." In: *Proceedings of the International Conference on Automated Planning and Scheduling* 30, pp. 143–151. DOI: 10.1609/icaps.v30i1.6655. URL: https://ojs.aaai.org/index.php/ICAPS/article/view/6655 (visited on 10/01/2022).

[26] UTC (2023). "Sphère de Bloch Interactive." URL: https://www.utc.fr/wschon/sr06/demonstrateur-algorithmes-quantiques-master/website/_site/qubitAnimation.html (visited on 07/05/2023).

[27] Wigderson, Avi (May 1982). "A New Approximate Graph Coloring Algorithm." In: *Proceedings of the Fourteenth Annual ACM Symposium on Theory of Computing*. STOC'82. New York, NY: Association for Computing Machinery, pp. 325–329. DOI: 10.1145800070.802207. URL: https://doi.org/10.1145/800070.802207 (visited on 01/10/2022).

[28] Wikipedia (May 2023). *Grover's algorithm*. en. Version ID: 1157281681. URL: https://en.wikipedia.org/w/index.php?title=Grover%27s_algorithmoldid=1157281681 (visited on 07/03/2023).

[29] Zhou, Zhaoyang et al. (Nov. 2014). "An Exact Algorithm with Learning for the Graph Coloring Problem." In: *Computers Operations Research* 51, pp. 282–301. DOI: 10.1016/j.cor.2014.05.017. URL: https://linkinghub.elsevier.com/retrieve/pii/S0305054814001464 (visited on 10/16/2022).

Index

Printed in the United States
by Baker & Taylor Publisher Services